The ESSENTIALS® of

Statistics I

Emil G. Milewski, Ph.D.

This book covers the usual course outline of
Statistics I. For more advanced topics, see
"THE ESSENTIALS OF STATISTICS II."

Research & Education Association
61 Ethel Road West
Piscataway, New Jersey 08854

THE ESSENTIALS®
OF STATISTICS I

Year 2000 Printing

Printed in the United States of America

Library of Congress Catalog Card Number 98-68338

International Standard Book Number 0-87891-658-X

ESSENTIALS is a registered trademark of
Research & Education Association, Piscataway, New Jersey 08854

WHAT "THE ESSENTIALS" WILL DO FOR YOU

This book is a review and study guide. It is comprehensive and it is concise.

It helps in preparing for exams, in doing homework, and remains a handy reference source at all times.

It condenses the vast amount of detail characteristic of the subject matter and summarizes the **essentials** of the field.

It will thus save hours of study and preparation time.

The book provides quick access to the important facts, principles, theorems, concepts, and equations in the field.

Materials needed for exams can be reviewed in summary form – eliminating the need to read and re-read many pages of textbook and class notes. The summaries will even tend to bring detail to mind that had been previously read or noted.

This "ESSENTIALS" book has been prepared by an expert in the field, and has been carefully reviewed to assure accuracy and maximum usefulness.

Dr. Max Fogiel
Program Director

CONTENTS

CHAPTER 1

INTRODUCTION

1.1 WHAT IS STATISTICS?

Statistics is the science of assembling, organizing, and analyzing data, as well as drawing conclusions about what the data means. Often these conclusions come in the form of predictions. Hence, the importance of statistics.

The first task of a statistician is to define the population and to choose the sample.

In most cases it is impractical or impossible to examine the entire group, which is called the population or universe. This being the case, one must instead examine a part of the population called a sample. The chosen sample has to reflect as closely as possible the characteristics of the population. For example, statistical methods were used in testing the Salk vaccine, which protects children against polio. The sample consisted of 400,000 children. Half of the children, chosen at random, received the Salk vaccine; the other half received an inactive solution. After the study was completed it turned out that 50 cases of polio appeared in the group which received the vaccine and 150 cases appeared in the group which did not receive the vaccine.

Based on that study of a sample of 400,000 children it was decided that the Salk vaccine was effective for the entire population.

A population can be finite or infinite.

EXAMPLE:

The population of all television sets in the USA is finite.

The population consisting of all possible outcomes in successive tosses of a coin is infinite.

Descriptive or deductive statistics collects data concerning a given group and analyzes it without drawing conclusions.

From an analysis of a sample, inductive statistics or statistical inference draws conclusions about the population.

1.2 VARIABLES

Variables will be denoted by symbols, such as x, y, z, t, a, M, etc. A variable can assume any of its prescribed values. A set of all the possible values of a variable is called its domain.

EXAMPLE:

For a toss of a die the domain is $\{1, 2, 3, 4, 5, 6\}$.

The variable can be discrete or continuous.

EXAMPLE:

The income of an individual is a discrete variable. It can be $1,000 or $1,000.01 but it cannot be between these two numbers.

EXAMPLE:

The height of a person can be 70 inches or 70.1 inches. It can also assume any value between these two numbers. Hence, height is a continuous variable.

Similarly, we are dealing with discrete data or continuous data. Usually, countings and enumerations yield discrete data, while measurements yield continuous data.

1.3 FUNCTIONS

Y is a function of X if for each value of X there corresponds one and only one value of Y. We write

$$Y = f(X)$$

to indicate that Y is a function of X. The variable X is called the independent variable and the variable Y is called the dependent variable.

EXAMPLE:

The distance s travelled by a car moving with a constant speed is a function of time t.

$$s = s(t)$$

We will also be dealing with the functions of two or more independent variables.

$$z = z(x, y)$$
$$y = f(x_1, x_2, \ldots, x_n)$$

The functional dependence can be depicted in the form of a table or an equation.

EXAMPLE:

Mr. Brown can present income from his real estate in the form of a table.

3

Year	Income
1979	18,000
1980	17,550
1981	17,900
1982	18,200
1983	18,600
1984	19,400
1985	23,500
1986	28,000

EXAMPLE:

$$y = 3x + 2$$

The value of a variable y is determined by the above equation.

x	-3	-2	-1	0	1	2	3
y	-7	-4	-1	2	5	8	11

We will be using rectangular coordinates.

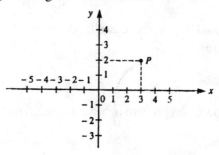

4

The coordinates of point P are $x = 3$ and $y = 2$. We write $(3, 2)$ to represent this.

Graphs illustrate the dependence between variables. We shall discuss linear graphs, bar graphs, and pie graphs.

1.4 DATA DESCRIPTION: GRAPHS

Repeated measurements yield data, which must be organized according to some principle. The data should be arranged in such a way that each observation can fall into one and only one category. A simple graphical method of presenting data is the pie chart, which is a circle divided into parts which represent categories.

EXAMPLE:

1979 Budget

38% came from individual income taxes
28% from social insurance receipts
13% from corporate income taxes
12% from borrowing
 5% from excise taxes
 4% other

This data can also be presented in the form of a bar chart or bar graph.

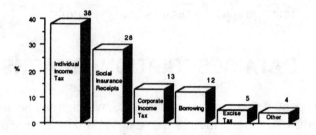

EXAMPLE:

The population of the United States for the years 1860, 1870, ..., 1950, 1960 is shown in the table below,

Year	1860	1870	1880	1890	1900	1910
Population in millions	31.4	39.8	50.2	62.9	76.0	92.0
Year	1920	1930	1940	1950	1960	
Population in millions	105.7	122.8	131.7	151.1	179.3	

and in this graph,

6

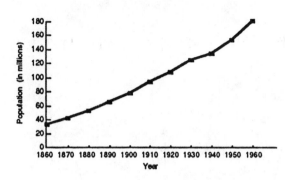

and in this bar chart.

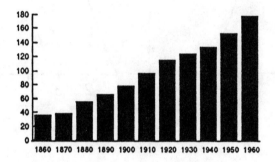

EXAMPLE:

A quadratic function is given by

$$y = x^2 + x - 2$$

We compute the values of y corresponding to various values of x.

x	-3	-2	-1	0	1	2	3
y	4	0	-2	-2	0	4	10

From this table the points of the graph are obtained:

7

$(-3, 4)$ $(-2, 0)$ $(-1, -2)$ $(0, -2)$ $(1, 0)$ $(2, 4)$ $(3, 10)$

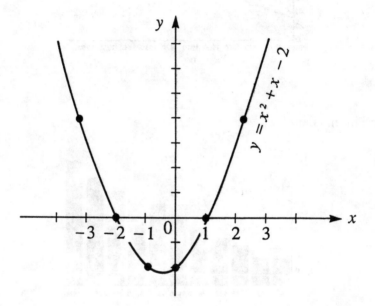

The curve shown is called a parabola. The general equation for a parabola is

$$y = ax^2 + bx + c, \quad a \neq 0$$

where a, b, and c are constants.

CHAPTER 2

FREQUENCY DISTRIBUTIONS

2.1 CLASS INTERVALS AND CLASS LIMITS

A set of measurements which has not been organized numerically is called raw data. An arrangement of raw numerical data in descending or ascending order of magnitude is called an array. The difference between the largest and smallest numbers in a set of data is called the range of the data.

EXAMPLE:

One hundred families were chosen at random and their yearly income was recorded.

Table 1
Income of 100 Families

Income in Thousands	Number of Families
10 – 14	3
15 – 19	12
20 – 24	19
25 – 29	20
30 – 34	23
35 – 39	18
40 – 44	5

Total   100

The range of the measurements is divided by the number of class intervals desired.

In this case we have seven classes. The number of individuals belonging to each class is called the class frequency.

For example, the class frequency of the class 35 – 39 is 18. The list of classes together with their class frequencies is called a frequency table or frequency distribution.

Table 1 is a frequency distribution of the income of 100 families. In Table 1 the labels 10 – 14 , 15 – 19, ..., 40 – 44 are called class intervals. The numbers 14 and 10 are called class limits; 10 is the lower class limit and 14 is the upper class limit.

In some cases, open class intervals are used, for example, "40 thousand and over." In Table 1 income is recorded to the nearest thousand. Hence, the class interval 30 – 34 includes all incomes from 29,500 to 34,499.

The exact numbers 29,500 and 34,499 are called true class limits or class boundaries. The smaller is the lower class boundary and the larger is the upper class boundary.

Class boundaries are often used to describe the classes. The difference between the lower and upper class boundaries is called the class width or class size. Usually, all class intervals of a frequency distribution have equal widths.

The class mark (or class midpoint) is the midpoint of the class interval.

$$\text{Class Midpoint} = \frac{\text{Lower Class Limit} + \text{Upper Class Limit}}{2}$$

The class mark of the class interval 35 – 39 is

$$\frac{35 + 39}{2} = 37$$

For purposes of mathematical analysis, all data belonging to a given class interval are assumed to coincide with the class mark.

**Guidelines for Constructing Class Intervals
and Frequency Distributions**

1. Find the range of the measurements, which is the difference between the largest and the smallest measurements.

2. Divide the range of the measurements by the approximate number of class intervals desired. The number of class intervals is usually between 5 and 20, depending on the data.

    Then round the result to a convenient unit, which should be easy to work with. This unit is a common width for the class intervals.

3. The first class interval should contain the smallest measurement and the last class interval should contain the largest measurement. No measurement should fall on a point of division between two intervals.

4. Determine the number of measurements which fall into each class interval; that is, find the class frequencies.

EXAMPLE:

The 25 measurements given below represent the sulphur level in the air for a sample of 25 days. The units used are parts per million.

Table 2

27	32	28	32	31
35	28	44	45	36
33	40	41	36	35
39	37	39	37	44
41	41	35	35	33

The lowest reading is 27 and the highest is 45.

Thus, the range is $45 - 27 = 18$. It will be convenient to use 5 class intervals $\frac{18}{5} = 3.6$. We round 3.6 to 4. The width for the class intervals is 4. The class intervals are labeled as follows: 26 – 29, 30 – 33, 34 – 37, 38 – 41, and 42 – 45.

Now, we can construct the frequency table and compute the class frequency for each class. The relative frequency of a class is defined as the frequency of the class divided by the total number of measurements. Table 3 shows the classes, frequencies and relative frequencies of the data in Table 2.

Table 3

Class Interval	Frequency	Relative Frequency
26 – 29	3	0.12
30 – 33	5	0.20
34 – 37	8	0.32
38 – 41	6	0.24
42 – 45	3	0.12
Total	25	1.00

2.2 FREQUENCY HISTOGRAMS AND RELATIVE FREQUENCY HISTOGRAMS

Polygons

A frequency histogram (or histogram) is a set of rectangles placed in the coordinate system.

The vertical axis is labeled with the frequencies and the horizontal axis is labeled with the class intervals. Over each class interval a rectangle is drawn with a height such that the area of the rectangle is proportional to the class frequency.

Often the height is numerically equal to the class frequency. Based on the results of Table 3, we obtain the histogram shown below.

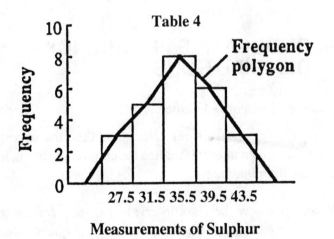

Measurements of Sulphur

A frequency polygon is obtained by connecting midpoints of the tops of the rectangles of the histogram.

The area bounded by the frequency polygon and the x–axis is equal to the sum of the areas of the rectangles in the histogram.

13

The relative frequency histogram is similar to the frequency histogram. Here, the vertical axis shows relative frequency. A rectangle is constructed over each class interval with a height equal to the class relative frequency. Based on Table 3, we obtain the relative frequency histogram shown in Table 5.

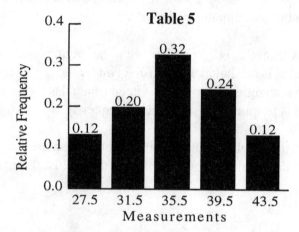

2.3 CUMULATIVE FREQUENCY DISTRIBUTIONS

Definition of Cumulative Frequency

The total frequency of all values less than the upper class boundary of a given interval is called the cumulative frequency up to and including that interval.

When we know the class intervals and their corresponding frequencies we can compute the cumulative frequency distribution.

Consider the class intervals and frequencies contained in Table 3, from which we compute the cumulative frequency distribution. See Table 6.

14

Table 6

Parts of Sulphur	Number of Days
Less than 26	0
Less than 30	3
Less than 34	8
Less than 38	16
Less than 42	22
Less than 46	25

Using the coordinate system we can present the cumulative frequency distribution graphically. Such a graph is called a cumulative frequency polygon or ogive. The ogive obtained from Table 6 is shown in Table 7.

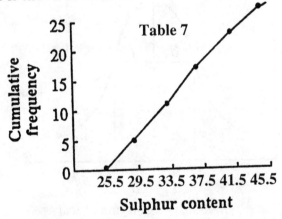

For a very large number of observations it is possible to choose very small class intervals and still have a significant number of observations falling within each class.

The sides of the frequency polygon or relative frequency polygon get smaller as class intervals get smaller. Such polygons closely approximate curves. Such curves are called frequency curves or relative frequency curves. Usually frequency curves can be obtained by increasing the number of class intervals, which requires a larger sample.

2.4 TYPES OF FREQUENCY CURVES

In applications we find that most of the frequency curves fall within one of the categories listed below.

1. One of the most popular is the bell-shaped or symmetrical frequency curve.

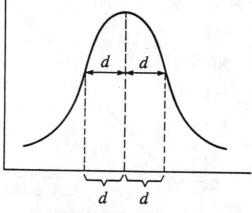

Bell-shaped or Symmetrical

Note that observations equally distant from the maximum have the same frequency. The normal curve has a symmetrical frequency curve.

2. The U–shaped curve has maxima at both ends.

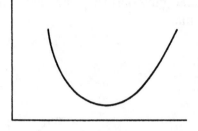

U-Shaped

3. A curve skewed to one side. The slope to the right of the maximum is steeper than the slope to the left. The opposite holds for the frequency curve skewed to the right.

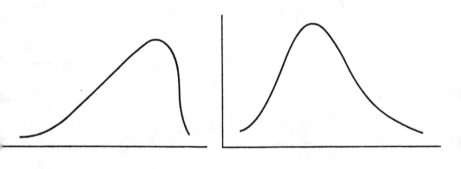

Skewed to the left   Skewed to the right
(negative skew)   (positive skew)

4. A J–shaped curve has a maximum at one end.

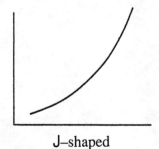

J–shaped

4. A multimodal (bimodal) frequency curve has more than one (two) maxima.

Multimodal

CHAPTER 3

NUMERICAL METHODS OF DESCRIBING DATA

3.1 INTRODUCTION

Once we have a sufficient number of measurements it is easy to find the frequency distribution. Graphic methods, however, are very often impractical or difficult to convey.

To remedy the situation, one can use a few numbers that describe the frequency distribution without the drawing of a real graph. Such numbers are called numerical descriptive measures. Each one describes a certain aspect of the frequency distribution. None of them yields the exact shape of the frequency distribution. Rather, they give us some notion of the general shape of the whole graph or parts of it.

For example, saying that somebody is 6' 4" and weighs 250 lbs. does not describe the person in detail but gives us the general idea of a stout man.

It is important to describe the center of the distribution of measurements as well as how the measurements behave about the center of the distribution. For that purpose we define central tendency and variability. In practical applications we deal with one of two essentially different situations:

1. The measurements are gathered about the whole population.

Numerical descriptive measures for a population are called parameters.

2. The measurements are gathered about the sample. Numerical descriptive measures for a sample are called statistics.

If we only have statistics, we are not able to calculate the values of parameters. But, using statistics, we can make reasonable estimates of parameters which describe the whole population. The most popular mathematical means of describing frequency distribution is an average. An average is a value which is representative or typical of a set of measurements.

Usually averages are called measures of central tendency. We will be using different kinds of averages, such as the arithmetic mean, the geometric mean and the harmonic mean. A different one should be applied depending on the data, the purpose and the required accuracy.

3.2 NOTATION AND DEFINITIONS OF MEANS

By

$$x_1, x_2, ..., x_n$$

we denote the measurements observed in a sample of size n. The letter i in x_i is called a subscript or index. It stands for any of the numbers $1, 2, ..., n$.

Throughout the book we will be using the summation notation. The symbol

$$\sum_{i=1}^{n} x_i$$

denotes the sum of all x_i's, that is,

$$\sum_{i=1}^{n} x_i = x_1 + x_2 + \ldots + x_{n-1} + x_n$$

EXAMPLE:

$$\sum_{i=1}^{4} x_i y_i = x_1 y_1 + x_2 y_2 + x_3 y_3 + x_4 y_4$$

EXAMPLE:

Let a be a constant. Then

$$\sum_{k=1}^{n} a x_k = a x_1 + a x_2 + \ldots + a x_n$$

$$= a(x_1 + x_2 + \ldots + x_n)$$

$$= a \sum_{k=1}^{n} x_k$$

In general

$$\sum a x_k = a \sum x_k$$

and

$$\sum (ax + by) = a \sum x + b \sum y$$

Often, when no confusion can arise we write $\sum_{k} x_k$ instead of $\sum_{k=1}^{n} x_k$.

The arithmetic mean, or mean, of a set of measurements is the sum of the measurements divided by the total number of measurements.

The arithmetic mean of a set of numbers x_1, x_2, ..., x_n is denoted by $\bar{x}$ (read "x bar").

$$\bar{x} = \frac{\sum_{i=1}^{n} x_i}{n} = \frac{x_1 + x_2 + ... + x_n}{n}$$

EXAMPLE:

The arithmetic mean of the numbers 3, 7, 1, 24, 11, 32 is

$$\bar{x} = \frac{3 + 7 + 1 + 24 + 11 + 32}{6} = 13$$

EXAMPLE:

Let f_1, f_2, ..., f_n be the frequencies of the numbers x_1, x_2, ..., x_n (i.e., number x_i occurs f_i times). The arithmetic mean is

$$\bar{x} = \frac{f_1 x_1 + f_2 x_2 + ... + f_n x_n}{f_1 + f_2 + ... + f_n} = \frac{\sum_{i=1}^{n} f_i x_i}{\sum_{i=1}^{n} f_i}$$

$$= \frac{\sum fx}{\sum f}.$$

Note that the total frequency, that is, the total number of cases, is $\sum_{i=1}^{n} f_i$.

EXAMPLE:

If the measurements 3, 7, 2, 8, 0, 4 occur with frequencies 3, 2, 1, 5, 10, and 6 respectively, then the arithmetic mean is

$$\bar{x} = \frac{3 \cdot 3 + 7 \cdot 2 + 2 \cdot 1 + 8 \cdot 5 + 0 \cdot 10 + 4 \cdot 6}{3 + 2 + 1 + 5 + 10 + 6} \approx 3.3$$

Keep in mind that the arithmetic mean is strongly affected by extreme values.

EXAMPLE:

Consider four workers whose annual salaries are $2,500, $3,200, $3,700, $48,000. The arithmetic mean of their salaries is

$$\frac{\$57,400}{4} = \$14,350$$

The figure $14,350 can hardly represent the typical annual salary of the four workers.

EXAMPLE:

The deviation d_i of x_i from its mean $\bar{x}$ is defined to be

$$d_i = x_i - \bar{x}$$

The sum of the deviations of $x_1, x_2, ..., x_n$ from their mean $\bar{x}$ is equal to zero. Indeed,

$$\sum_{i=1}^{n} d_i = \sum_{i=1}^{n} (x_i - \bar{x}) = 0$$

Thus,

$$\sum_{i=1}^{n} (x_i - \bar{x}) = \sum_{i=1}^{n} x_i - n\bar{x} = \sum x_i - n\frac{\sum x_i}{n}$$

$$= \sum x_i - \sum x_i = 0 .$$

EXAMPLE:

If $z_1 = x_1 + y_1 , \ldots , z_n = x_n + y_n$, then $\bar{z} = \bar{x} + \bar{y}$.

Indeed

$$\bar{x} = \frac{\sum x}{n} , \ \bar{y} = \frac{\sum y}{n} , \ \text{and} \ \bar{z} = \frac{\sum z}{n} .$$

We have

$$\bar{z} = \frac{\sum z}{n} = \frac{\sum (x + y)}{n} = \frac{\sum x}{n} + \frac{\sum y}{n} = \bar{x} + \bar{y} .$$

EXAMPLE:

Consider the numbers $x_1, x_2, \ldots, x_n$ whose deviations from any number a are $d_1, d_2, \ldots, d_n$ respectively. Then

$$\bar{x} = a + \bar{d} .$$

The deviation d_i of x_i from a is defined to be

$$d_i = x_i - a .$$

Hence

$$\bar{x} = \frac{\sum x_i}{n} = \frac{\sum (d_i + a)}{n} = \frac{na}{n} + \frac{\sum d_i}{n} = a + \bar{d} .$$

Similarly, if $x_1, x_2, \ldots, x_n$ have frequencies $f_1, f_2, \ldots, f_n$ respectively, then

$$\bar{x} = a + \frac{\sum f_i d_i}{N} , \ \text{where} \ N = \sum_{i=1}^{n} f_i$$

Indeed,

$$\bar{x} = \frac{\sum f_i x_i}{\sum f_i} = \frac{\sum f_i x_i}{N} = \frac{\sum f_i (a + d_i)}{N}$$

$$= \frac{a \sum f_i}{N} + \frac{\sum f_i d_i}{N} = a + \frac{\sum f_i d_i}{N}.$$

The arithmetic mean plays an important role in statistical infer-ence.

We will be using different symbols for the sample mean and the population mean. The population mean is denoted by μ, and the sample mean is denoted by $\bar{x}$. The sample mean $\bar{x}$ will be used to make inferences about the corresponding population mean μ.

EXAMPLE:

Suppose a bank has 500 savings accounts. We pick a sample of 12 accounts. The balance on each account is

$657	$284	$ 51
215	73	327
65	412	218
539	225	195

The sample mean $\bar{x}$ is

$$\bar{x} = \frac{\sum_{i=1}^{12} x_i}{12} = \$271.75$$

The average amount of money in each of 12 sampled accounts is $271.75. Using this information we estimate the total amount of money in the bank to be

25

$$\$271.75 \cdot 500 = \$135,875.$$

Definition of Geometric Mean

The geometric mean g (or G) of a set of n numbers

$$x_1, x_2, \ldots, x_n$$

is the nth root of the product of the numbers

$$g = \sqrt[n]{x_i \cdot x_2 \cdot \ldots \cdot x_n}$$

EXAMPLE:

The geometric mean of the numbers 3, 9, 27 is

$$g = \sqrt[3]{3 \cdot 9 \cdot 27} = 9 .$$

EXAMPLE:

Find the geometric mean of the numbers 3, 5, 7, 8, 10, 13, 16.
We begin with

$$g = \sqrt[7]{3 \cdot 5 \cdot 7 \cdot 8 \cdot 10 \cdot 13 \cdot 16} = \sqrt[7]{1,747,200}$$

Using common logarithms

$$\log g = \frac{1}{7}\log 1,747,200 = \frac{1}{7} \cdot 6.2423 = 0.8918$$

Hence,

$$g = 7.794$$

EXAMPLE:

The frequencies of measurements $x_1, x_2, \ldots, x_n$ are $f_1, f_2, \ldots, f_n$ respectively. The geometric mean of the measurements is

$$g = \sqrt[N]{\underbrace{x_1 \cdot x_1 \cdot \ldots \cdot x_1}_{f_1 \text{ times}} \cdot \underbrace{x_2 \cdot x_2 \cdot \ldots \cdot x_2}_{f_2 \text{ times}} \cdot \ldots \cdot \underbrace{x_n \cdot \ldots \cdot x_n}_{f_n \text{ times}}}$$

$$= \sqrt[N]{x_1^{f_1} \cdot x_2^{f_2} \cdot x_n^{f_n}} \quad \text{where } N = \sum_{1}^{n} f_i$$

If all the numbers are positive, then

$$\log g = \frac{1}{N} \log \left(x_1^{f_1} \cdot x_2^{f_2} \cdot \ldots \cdot x_n^{f_n} \right)$$

$$= \frac{1}{N} \left(f_1 \log x_1 + \ldots + f_n \log x_n \right)$$

$$= \frac{\sum f_i \log x_i}{\sum f_i} .$$

EXAMPLE:

The money deposited in an interest-bearing account increased from \$1,000 to \$5,000 in three years. What was the average percentage increase per year?

The increase is 500%. But the average percentage increase is not $\dfrac{500\%}{3}$.

Denote the average percentage increase by q. Then

amount deposited	1,000
amount after first year	$1{,}000 + q1{,}000 = 1{,}000(1 + q)$

amount after second year	$1,000(1 + q) + 1,000(1 + q)q$
	$= 1,000(1 + q)^2$
amount after third year	$= 1,000(1 + q)^2 + 1,100(1 + q^2)q$
	$= 1,000(1 + q)^3$
	$= 5,000$

Hence

$$(1 + q)^3 = 5$$

and

$$q = \sqrt[3]{5} - 1$$

In general, if the initial amount is A and the yearly interest is q, then after n years, the amount is M.

$$M = A(1 + q)^n$$

This equation is called the compound interest formula.

Definition of Weighted Arithmetic Mean

With the numbers $x_1, x_2, ..., x_n$ we associate weighting factors or weights $w_1, w_2, ..., w_n$ depending on how significant each number is. The weighted arithmetic mean is defined by

$$\bar{x} = \frac{x_1 w_1 + x_2 w_2 + ... + x_n w_n}{w_1 + w_2 + ... + w_n} = \frac{\sum x_n w_n}{\sum w_n}$$

EXAMPLE:

The pilot has to pass three tests. The second test is weighted three times as much as the first and the third test is weighted four times as much as the first. The pilot reached the score of 40 on the first test, 45 on the second test and 60 on the third test.

The weighted mean is

$$\frac{40 \cdot 1 + 45 \cdot 3 + 60 \cdot 4}{8} = 51.875$$

EXAMPLE:

If f_1 numbers have mean m_1, f_2 numbers have mean m_2, ..., f_n numbers have mean m_n, then the mean of all numbers is

$$\bar{x} = \frac{f_1 m_1 + f_2 m_2 + \ldots + f_n m_n}{f_1 + f_2 + \ldots + f_n}$$

Definition of Harmonic Mean

The harmonic mean h of numbers $x_1, x_2, \ldots, x_n$ is the reciprocal of the arithmetic mean of the reciprocals of the numbers:

$$h = \frac{1}{\frac{1}{n} \sum_{i=1}^{n} \frac{1}{x_i}} = \frac{n}{\sum \frac{1}{x_i}}$$

EXAMPLE:

The harmonic mean of the numbers 2, 4, 9 is

$$h = \frac{3}{\frac{1}{2} + \frac{1}{4} + \frac{1}{9}} = 3.48$$

EXAMPLE:

During four successive years the prices of gasoline were 70, 75, 78 and 95 cents per gallon. Find the average cost of gasoline over the four year period. There are two methods of computing this average.

Method 1.

Suppose the car owner used 100 gallons each year. Then

$$\text{Average Cost} = \frac{\text{total cost}}{\text{total number of gallons}}$$

$$= \frac{0.7 \cdot 100 + 0.75 \cdot 100 + 0.78 \cdot 100 + 0.95 \cdot 100}{400}$$

$$= 0.795 \ \$/\text{gal} \ .$$

Method 2.

Suppose the car owner spends \$100 on gasoline each year. Then

$$\text{Average Cost} = \frac{\text{total cost}}{\text{total number of gallons}}$$

$$= \frac{400}{509.66} = 0.785 \ \$/\text{gal} \ .$$

Both averages are correct. Depending on assumed conditions we obtain different answers.

Let g be the geometric mean of a set of positive numbers x_1, $x_2, \ldots, x_n$ and let $\bar{x}$ be the arithmetic mean and h the harmonic mean. Then

$$h \leq g \leq \bar{x}$$

The equality holds only if all the numbers $x_1, x_2, \ldots, x_n$ are identical.

Definition of Root Mean Square

The root mean square or quadratic mean of the numbers $x_1, x_2, \ldots, x_n$ is denoted by $\sqrt{\overline{x^2}}$ and defined as

$$\sqrt{\overline{x^2}} = \sqrt{\frac{\sum\limits_{i=1}^{n} x_i^2}{n}}$$

EXAMPLE:

The quadratic mean of the numbers 2, 3, 5, 7 is

$$\sqrt{\overline{x^2}} = \sqrt{\frac{2^2 + 3^2 + 5^2 + 7^2}{4}} = 4.66$$

EXAMPLE:

The quadratic mean of two positive numbers, a and b, is not smaller than their geometric mean.

$$\sqrt{ab} \leq \sqrt{\frac{a^2 + b^2}{2}}$$

Indeed, since

$$0 \leq (a - b)^2 = a^2 - 2ab + b^2$$

we obtain

$$ab \leq \frac{a^2 + b^2}{2}$$

and

$$\sqrt{ab} \leq \sqrt{\frac{a^2 + b^2}{2}}$$

All means can also be computed for the grouped data, that is, when data are presented in a frequency distribution. Then all values within a given class interval are considered to be equal to the class mark or midpoint of the interval.

3.3 MEASURES OF CENTRAL TENDENCY

Definition of The Mode

The mode of a set of numbers is that value which occurs most often (with the highest frequency).

Observe that the mode may not exist. Also, if the mode exists it may not be unique. For example, for the numbers 1, 1, 2, 2 the mode is not unique.

EXAMPLE:

The set of numbers 2, 2, 4, 7, 9, 9, 13, 13, 13, 26, 29 has mode 13.

The set of numbers which has two or more modes is called bimodal.

For grouped data – data presented in the form of a frequency table – we do not know the actual measurements, but only how many measurements fall into each interval. In such a case the mode is the midpoint of the class interval with the highest frequency.

Note that the mode can also measure popularity. In this sense we can determine the most popular model of car or the most popular actor.

EXAMPLE:

One can compute the mode from a histogram or frequency distribution.

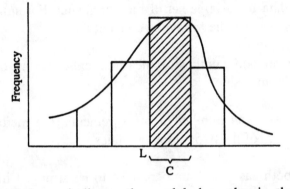

The shaded area indicates the modal class, that is, the class containing the mode.

$$\text{Mode} = L + c\left(\frac{\Delta_1}{\Delta_1 + \Delta_2}\right)$$

where
    L is lower class boundary of modal class
    c is the size of modal class interval
    Δ_1 is the excess of the modal frequency over the frequency of next lower class
    Δ_2 is the excess of the modal frequency over the frequency of next higher class.

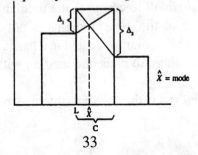

33

Definition of Median

The median of a set of numbers is defined to be the middle value when the numbers are arranged in order of magnitude.

Usually the median is used to measure the midpoint of a large set of numbers. For example, we can talk about the median age of people getting married. Here the median reflects the central value of the data for a large set of measurements. For small sets of numbers we use the following convention:

— For an odd number of measurements, the median is the midpoint value.

— For an even number of measurements, the median is the average of the two midpoint values.

In both cases the numbers have to be arranged in order of magnitude.

EXAMPLE:

The scores of a test are 78, 79, 83, 83, 87, 92, 95. Hence, the median is 83.

EXAMPLE:

The median of the set of numbers 21, 25, 29, 33, 44, 47 is

$$\frac{29 + 33}{2} = 31 \ .$$

It is more difficult to compute the median for grouped data. The exact value of the measurements is not known, hence we know only that the median is located in a particular class interval. The problem is where to place the median within this interval.

For grouped data, the median obtained by interpolation, is

given by

$$\text{Median} = L + \frac{c}{f_{\text{median}}} \left(\frac{n}{2} - (\Sigma f)_{\text{cum}} \right)$$

where L =   lower class limit of the interval that contains the median

            c =   size of median class interval

       f_{median} =   frequency of median class

          n =   total frequency

    $(\Sigma f)_{\text{cum}}$ =   the sum of frequencies (cumulative frequency) for all classes before the median class

EXAMPLE:

The weight of 50 men is depicted in the table below in the form of frequency distribution.

Weight	Frequency
115 – 121	2
122 – 128	3
129 – 135	13
136 – 142	15
143 – 149	9
150 – 156	5
157 – 163	3
Total	50

Class 136 – 142 has the highest frequency.

The mode is the midpoint of the class interval with the highest frequency.

$$\text{Mode} = \frac{135.5 + 142.5}{2} = 139$$

We can also use the formula

$$\text{Mode} = L + c\left(\frac{\Delta_1}{\Delta_1 + \Delta_2}\right)$$

here

$L \quad = \quad 135.5$

$c \quad = \quad 7$

$\Delta_1 \quad = \quad 2 \ (15 - 13 = 2)$

$\Delta_2 \quad = \quad 6 \ (15 - 9 = 6)$

$$\text{Mode} = 135.5 + 7 \cdot \frac{2}{2 + 6} = 137.25$$

The median is located in class $136 - 142$.

We have

$$\text{Median} = L + \frac{c}{f_{\text{median}}}\left(\frac{n}{2} - (\Sigma f)_{\text{cum}}\right)$$

$L \quad = \quad 135.5$

$c \quad = \quad 7$

$f_{\text{median}} \quad = \quad 15$

$n \quad = \quad 50$

$(\Sigma f)_{\text{cum}} \quad = \quad 2 + 3 + 13 = 18$

Hence

$$\text{Median} = 135.5 + \frac{7}{15}\left(\frac{50}{2} - 18\right) = 138.77$$

To compute the arithmetic mean for grouped data we compute midpoint x_i of each of the intervals and use the formula

$$\bar{x} = \frac{\displaystyle\sum_{i=1}^{n} f_i x_i}{\displaystyle\sum_{i=1}^{n} f_i}$$

We have

$$\bar{x} = \frac{118 \cdot 2 + 125 \cdot 3 + 132 \cdot 13 + 139 \cdot 15 + 146 \cdot 9 + 153 \cdot 5 + 160 \cdot 3}{50}$$

$$= 139.42$$

For unimodal frequency curves which are moderately asymmetrical, the following empirical relation holds:

$$\boxed{\frac{\text{Mean} - \text{Mode}}{3} = \text{Mean} - \text{Median}}$$

For symmetrical curves the mean, mode and median all coincide.

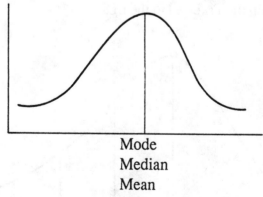

Mode
Median
Mean

For skewed distributions we have

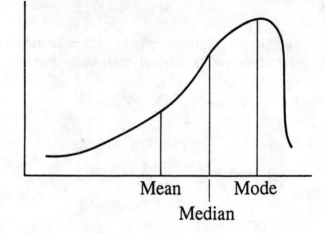

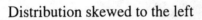

Distribution skewed to the left

Distribution skewed to the right.

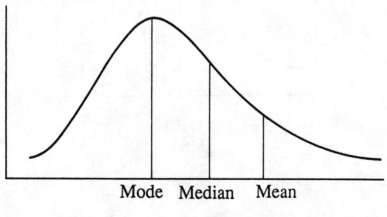

CHAPTER 4

MEASURES OF VARIABILITY

4.1 RANGE AND PERCENTILES

The degree to which numerical data tend to spread about an average value is called the variation or dispersion of the data. We shall define various measures of dispersion.

The simplest measure of data variation is the range.

Definition of Range

The range of a set of numbers is defined to be the difference between the largest and the smallest number of the set.

EXAMPLE:

The range of the numbers 3, 6, 21, 24, 38 is $38 - 3 = 35$.

For grouped data the range is the difference between the upper limit of the last interval and the lower limit of the first interval.

Next we shall define percentiles.

Definition of Percentiles

The nth percentile of a set of numbers arranged in order of magnitude is that value which has $n\%$ of the numbers below it and $(100 - n)\%$ above it.

EXAMPLE:

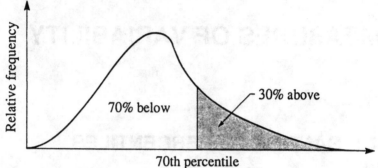

70% below

30% above

70th percentile

The 70th percentile of a set of numbers.

Percentiles are often used to describe the results of achievement tests. For example, someone graduates in the top 10% of his class. Frequently used percentiles are the 25th, 50th, and 75th percentiles, which are called the lower quartile, the middle quartile (median), and the upper quartile, respectively.

Definition of Interquartile Range

The interquartile range, denoted IQR, of a set of numbers is the difference between the upper and lower quartiles.

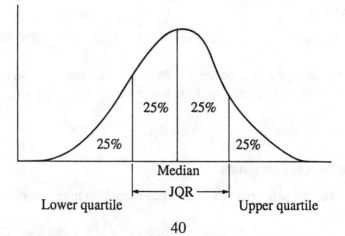

25%   25%

25%   25%

Median

JQR

Lower quartile   Upper quartile

Now we shall introduce an important concept of deviation.

The deviation of a number x from its mean $\overline{x}$ is defined to be

$$x - \overline{x}$$

Using deviations we can construct many different measures of variability.

Observe that the mean deviation for any set of measurements is always zero. Indeed, let $x_1, x_2, ..., x_n$ be measurements. Their mean is given by

$$\overline{x} = \frac{\sum x_i}{n}$$

The deviations are $x_1 - \overline{x}, x_2 - \overline{x}, ..., x_n - \overline{x}$ and their mean is equal to

$$\frac{\sum_{i=1}^{n} (x_i - \overline{x})}{n} = \frac{\sum x_i}{n} - \overline{x} = 0$$

4.2 MEASURES OF DISPERSION

Definition of Average Deviation

The average deviation of a set of n numbers $x_1, x_2, ..., x_n$ is defined by

$$\text{Average Deviation} = \frac{\sum_{i=1}^{n} |x_i - \overline{x}|}{n} = \overline{|x - \overline{x}|}$$

where $\overline{x}$ is the arithmetic mean of the numbers $x_1, x_2, ..., x_n$.

EXAMPLE:

We find the average deviation of the numbers 3, 5, 6, 8, 13, 21 by

$$\bar{x} = \frac{3 + 5 + 6 + 8 + 13 + 21}{6} = 9.33$$

$$\text{Average Deviation} = \frac{6.33 + 4.33 + 3.33 + 1.33 + 4.33 + 12.33}{6}$$

$$= 5.33$$

If the frequencies of the numbers $x_1, x_2, ..., x_n$ are $f_1, f_2, ..., f_n$ respectively, then the average deviation becomes

$$\text{Average Deviation} = \frac{\sum_i f_i \left| x_i - \bar{x} \right|}{\sum f_i}$$

We will be using this formula for grouped data where x_i's represent class marks and f_i represents class frequencies.

The sum $\sum_{i=1}^{n} \left| x_i - a \right|$ is the maximum when a is the median.

Definition of Standard Deviation

The standard deviation of a set $x_1, x_2, ..., x_n$ of n numbers is defined by

$$s = \sqrt{\frac{\sum_{i=1}^{n} (x_i - \bar{x})^2}{n}} = \sqrt{\overline{(x - \bar{x})^2}}$$

The sample standard deviation is denoted by s, while the corresponding population standard deviation is denoted by σ.

42

For grouped data we use the modified formula for standard deviation. Let the frequencies of the numbers $x_1, x_2, \ldots, x_n$ be $f_1, f_2, \ldots, f_n$ respectively. Then

$$s = \sqrt{\frac{\sum f_i (x_i - \overline{x})^2}{\sum f_i}} = \sqrt{\frac{\sum f(x - \overline{x})^2}{\sum f}}$$

Often, in the definition of the standard deviation the denominator is not n but $n-1$. For large values of n the difference between the two definitions is negligible.

Definition of Variance

The variance of a set of measurements is defined as the square of the standard deviation. Thus

$$s^2 = \frac{\sum_{i=1}^{n} (x_i - \overline{x})^2}{n}$$

or

$$s^2 = \frac{\sum_{i=1}^{n} f_i (x_i - \overline{x})^2}{\sum_{i=1}^{n} f_i}$$

Usually the variance of the sample is denoted by s^2 and the corresponding population variance is denoted by σ^2.

EXAMPLE:

A simple manual task was given to six children and the time each child took to complete the task was measured. Results are shown in the table.

x_i	$x_i - \overline{x}$	$(x_i - \overline{x})^2$
12	2.5	6.25
9	−0.5	0.25
11	1.5	2.25
6	−3.5	12.25
10	0.5	0.25
9	−0.5	0.25

Total 57   0   21.5

For this sample we shall find the standard deviation and variance.

The average $\overline{x}$ is 9.5.

$$\overline{x} = 9.5$$

The standard deviation is

$$s = \sqrt{\frac{21.5}{6}} = 1.893$$

and the variance is

$$s^2 = 3.583$$

4.3 SIMPLIFIED METHODS FOR COMPUTING THE STANDARD DEVIATION AND VARIANCE

By definition

$$s^2 = \frac{\sum_{i=1}^{n} (x_i - \bar{x})^2}{n}$$

Hence

$$s^2 = \frac{\sum (x - \bar{x})^2}{n} = \frac{\sum (x^2 - 2x\bar{x} + \bar{x}^2)}{n}$$

$$= \frac{\sum x^2}{n} - 2\bar{x} \frac{\sum x}{n} + \frac{n\bar{x}^2}{n} = \frac{\sum x^2}{n} - 2\bar{x}^2 + \bar{x}^2$$

$$= \bar{x}^2 - \bar{x}^2 = \frac{\sum x_i^2}{n} - \left(\frac{\sum x_i}{n}\right)^2$$

and

$$s = \sqrt{\bar{x}^2 - \bar{x}^2} = \sqrt{\frac{\sum x_i^2}{n} - \left(\frac{\sum x_i}{n}\right)^2}$$

Similarly for

$$s^2 = \frac{\sum_{i=1}^{n} f_i(x_i - \bar{x})^2}{\sum f_i}$$

We find

$$s^2 = \frac{\sum f(x - \bar{x})^2}{\sum f} = \frac{\sum fx^2 - 2\bar{x}\sum fx + \bar{x}^2\sum f}{\sum f}$$

$$= \frac{\sum fx^2}{\sum f} - \bar{x}^2 = \frac{\sum fx^2}{\sum f} - \left(\frac{\sum fx}{\sum f}\right)^2$$

EXAMPLE:

Consider the example in Section 4.2. We shall use the formula

$$s^2 = \overline{x^2} - \bar{x}^2$$

to compute the value of s^2.

$$s^2 = \frac{563}{6} - (9.5)^2 = 3.583$$

and

$$s = 1.893$$

Remember that $\overline{x^2}$ denotes the mean of the squares of all values of x and $\bar{x}^2$ denotes the square of the mean of all x's.

Let D be an arbitrary constant and

$$d_i = x_i - D$$

be the deviations of x_i from D. Then

$$x_i = d_i + D \quad \text{and}$$
$$\bar{x} = \bar{d} + D$$

Hence

$$s^2 = \overline{(x - \overline{x})^2} = \overline{(d - \overline{d})^2} = \overline{d^2 - 2\overline{d}d + \overline{d}^2}$$

$$= \overline{d^2} - \overline{d}^2 = \frac{\sum f\,d^2}{\sum f} - \left(\frac{\sum f\,d}{\sum f}\right)^2$$

EXAMPLE:

We find the standard deviation for the data shown below. The height of 100 students was measured and recorded.

Height (inches)	Class Mark x	x^2	Frequency f	fx^2	fx
60 – 62	61	3,721	7	26,047	427
63 – 65	64	4,096	21	86,016	1,344
66 – 68	67	4,489	37	166,093	2,479
69 – 71	70	4,900	26	127,400	1,820
72 – 74	73	5,329	9	47,961	657
			$\Sigma f =$ 100	$\Sigma fx^2 =$ 453,517	$\Sigma fx =$ 6,727

We apply the formula

$$s = \sqrt{\frac{\sum fx^2}{\sum f} = \left(\frac{\sum fx}{\sum f}\right)^2}$$

Hence

$$s = \sqrt{4,535 - 4,525} = 3.16$$

Using the empirical rule we can interpret the standard deviation of a set of measurements.

Empirical Rule

If a set of measurements has a bell-shaped histogram, then:

1. the interval $\bar{x} \pm s$ contains approximately 68% of the measurements.

2. $\bar{x} \pm 2s$ contains approximately 95% of the measurements.

3. $\bar{x} \pm 3s$ contains approximately all the measurements.

Note that from the empirical rule we can find the approximate value of the sample standard deviation s. Approximately 95% of all the measurements are located in the interval $\bar{x} \pm 2s$. The length of this interval is $4s$. Hence the range of the measurements is equal to approximately $4s$.

$$\text{Approximate Value of } s = \frac{\text{range}}{4}$$

For normal (bell-shaped) distributions we have

Empirical Rule

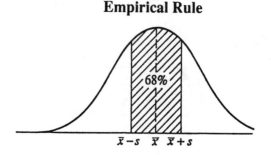

$$\bar{x} - s \quad \bar{x} \quad \bar{x} + s$$

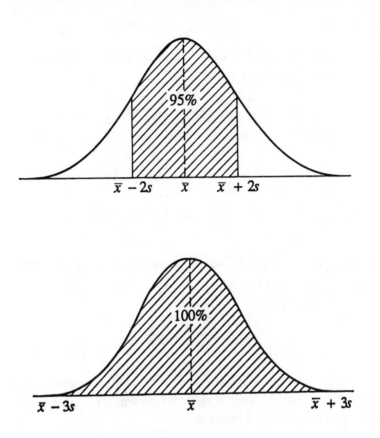

We shall discuss some properties of the standard deviation.

In the definition of the standard deviation

$$s = \sqrt{\frac{\sum_{i=1}^{n} (x_i - \bar{x})^2}{n}}$$

we can replace $\bar{x}$ by any other average y which is not the arithmetic mean

$$s = \sqrt{\frac{\sum (x_i - y)^2}{n}}$$

Note that of all such standard deviations, the one with $y = \overline{x}$ is the minimum. Indeed

$$\frac{\sum (x - y)^2}{n} = \frac{\sum (x^2 - 2xy + y^2)}{n}$$

$$= \frac{\sum x^2 - 2y \sum x + ny^2}{n}$$

$$= y^2 - 2y \frac{\sum x}{n} + \frac{\sum x^2}{n}$$

Equation $y^2 + ay + b$, where a and b are constants, has a minimum value if and only if $y = -\frac{1}{2} a$. Hence

$$y = \frac{\sum x}{n} = \overline{x}$$

Consider two sets consisting of N_1 and N_2 measurements or two frequency distributions with total frequencies N_1 and N_2. The variances are s_1^2 and s_2^2 respectively.

If both sets have the same mean $\overline{x}$, then the combined (or pooled) variance of both sets or both frequency distributions is given by

$$s^2 = \frac{N_1 s_1^2 + N_2 s_2^2}{N_1 + N_2}$$

4.4 CODING METHODS

We use the coding methods to simplify calculations. Data are generally coded by one or both of the following operations:

1. addition (or subtraction) of a constant A to (or from) each measurement.

2. multiplication (or division) of each measurement by a constant.

We shall describe the results of coding. Let $x_1, x_2, \ldots, x_n$ be n measurements with the arithmetic mean $\overline{x}$ and sample standard deviation s. Then

1. If a constant A was subtracted from each measurement, the mean and the standard deviation are given by

$$\overline{x}_A = \overline{x} - A$$

$$s_A = s$$

2. If each measurement was multiplied by a positive constant k, the mean and the standard deviation are given by

$$\overline{x}_k = k\overline{x}$$

$$s_k = k s$$

Next we shall describe the coding procedure for grouped data.

The frequency distribution is given with the class intervals of equal size c. Each class mark x_i is coded into a corresponding value y_i by

$$x_i = A + cy_i$$

where A and c are constants. The standard deviation is

$$s = \sqrt{\overline{(x - \bar{x})^2}}$$

but

$$x = A + cy$$

and

$$\bar{x} = A + c\bar{y}$$

$$x - \bar{x} = c(y - \bar{y}) \; .$$

Thus

$$s^2 = \overline{(x - \bar{x})^2} = \overline{c^2(y - \bar{y})^2}$$

$$= \overline{c^2(y^2 - 2y\bar{y} + \bar{y}^2)} = c^2(\overline{y^2} - 2\bar{y}^2 + \bar{y}^2)$$

$$= c^2(\overline{y^2} - \bar{y}^2)$$

and

$$\boxed{s = c\sqrt{\overline{y^2} - \bar{y}^2} = c\sqrt{\frac{\sum f y^2}{\sum f} - \left(\frac{\sum f y}{\sum f}\right)^2}}$$

Charlier's Check

We shall describe Charlier's check using an example.

EXAMPLE:

The results of the I.Q. (intelligence quotient) test of 150 students are shown in the table. The data are grouped and only class marks are recorded.

Class Mark	95	100	105	110	115	120
Frequency	3	5	9	14	17	21
Class Mark	125	130	135	140	145	150
Frequency	25	19	17	11	7	2

Using the coding method we find the mean and the standard deviation.

Class mark 125 has the highest frequency. We denote

$$A = 125$$

The class intervals are of equal size $k = 5$.

$$x_i = A + ky_i$$

For example

$$130 = 125 + 5\,(1) \quad \text{and}$$
$$110 = 125 + 5\,(-3)$$

For each x_i we compute y_i. The mean $\bar{x}$ can be calculated from the equation

$$\bar{x} = A + k\bar{y}$$

where

$$\bar{y} = \frac{\sum f_i y_i}{\sum f_i}$$

53

The standard deviation is given by

$$s = k\sqrt{\overline{y^2} - \overline{y}^2} = k\sqrt{\frac{\sum f y^2}{\sum f} - \left(\frac{\sum f y}{\sum f}\right)^2}$$

where

$$x_i = A + ky_i$$

The results for each interval are listed in the table below.

	x_i	y_i	f_i	f_iy_i	$f_iy_i^2$
	95	−6	3	−18	108
	100	−5	5	−25	125
	105	−4	9	−36	144
	110	−3	14	−42	126
	115	−2	17	−34	68
	120	−1	21	−21	21
$A =$	125	0	25	0	0
	130	1	19	19	19
	135	2	17	34	68
	140	3	11	33	99
	145	4	7	28	112
	150	5	2	10	50
			$\sum f_i =$ 150	$\sum f_iy_i =$ −52	$\sum f_iy_i^2 =$ 940

The mean is

$$\bar{x} = A + k\bar{y} = A + k \frac{\sum f_i y_i}{\sum f_i}$$

$$= 125 + 5 \cdot \frac{-52}{150} = 123.27$$

and the standard deviation is

$$s = k \sqrt{\frac{\sum f_i y_i^2}{\sum f_i} - \left(\frac{\sum f_i y_i}{\sum f_i}\right)^2}$$

$$= 5 \sqrt{\frac{940}{150} - \left(\frac{-52}{150}\right)^2} = 12.4$$

Charlier's check was the identity

$$\sum f_i(y_i + 1) = \sum f_i y_i + \sum f_i$$

to verify the mean and the identity

$$\sum f_i(y_i + 1)^2 = \sum f_i y_i^2 + 2\sum f_i y_i + \sum f_i$$

to verify the value of the standard deviation. The values of $f_i(y_i + 1)$ and of $f_i(y_i + 1)^2$ are given in the table below.

Also the sums $\Sigma f_i(y_i + 1)$ and $\Sigma f_i(y_i + 1)^2$ are computed.

We have

$$\Sigma f_i y_i = -52$$
$$\Sigma f_i = 150$$

and

$$\Sigma f_i(y_i + 1) = 98$$

Indeed

$$98 = -52 + 150$$

$y_i + 1$	f_i	$f_i(y_i + 1)$	$f_i(y_i + 1)^2$
–5	3	–15	75
–4	5	–20	80
–3	9	–27	81
–2	14	–28	56
–1	17	–17	17
0	21	0	0
1	25	25	25
2	19	38	76
3	17	51	153
4	11	44	176
5	7	35	175
6	2	12	72
	$\Sigma f_i =$ 150	$\Sigma f_i(y_i + 1) =$ 98	$\Sigma f_i(y_i + 1)^2 =$ 986

We have

$$\Sigma f(y + 1)^2 = 986$$
$$\Sigma f y^2 = 940$$
$$2\Sigma f y = -104$$
$$\Sigma f = 150$$

Indeed

$$986 = 940 + 2(-52) + 150.$$

Sheppard's Correction

The value of the standard deviation s and of variance s^2 depends on how the data are grouped into classes. The error which occurs is called the grouping error. To compensate for the grouping error the adjustment is introduced

$$\boxed{\text{Corrected Variance} = \text{Calculated Variance} - \frac{k^2}{12}}$$

where k is the class interval size. $\frac{k^2}{12}$ is called Sheppard's correction. When Sheppard's correction should be applied depends on the situation.

For the normal distribution the mean deviation is equal to 0.7979 times the standard deviation and the semi-interquartile range is equal to 0.6745 times the standard deviation. For the normal distribution we have the formulas

$$\text{Mean Deviation} = \frac{4}{5} \cdot \text{Standard Deviation}$$

$$\text{Semi} - \text{interquartile Range} = \frac{2}{3} \cdot \text{Standard Deviation}$$

Standardized Variable

We define the new variable

$$y = \frac{x - \bar{x}}{s}$$

which is called a standardized variable. It measures the deviation from the mean in units of the standard deviation. This variable is dimensionless, that is, it is independent of the units used.

Coefficient of Variation

We shall call the actual dispersion or variation the absolute dispersion. The absolute dispersion is determined from the standard deviation or any other measure of dispersion.

The same value of dispersion can have an entirely different meaning in different situations. This fact is taken into account in the relative dispersion

$$\text{Relative Dispersion} = \frac{\text{Absolute Dispersion}}{\text{Average}}$$

When the absolute dispersion is the standard deviation we obtain

$$\frac{s}{\bar{x}} = \text{Coefficient of Variation}$$

Observe that the coefficient of variation loses its meaning when $\bar{x}$ is close to zero.

CHAPTER 5

MOMENTS: PARAMETERS OF DISTRIBUTIONS

5.1 MOMENTS

The set of numbers $x_1, x_2, \ldots, x_n$ is given. Their sth moment is defined by

$$\overline{x^s} = \frac{x_1^s + x_2^s + \ldots + x_n^s}{n} = \frac{\sum\limits_{i=1}^{n} x_i^s}{n}$$

For $s = 1$ the first moment is the arithmetic mean $\overline{x}$.

The sth moment about the mean $\overline{x}$ is defined as

$$m_s = \frac{\sum\limits_{i=1}^{n} (x_i - \overline{x})^s}{n} = \overline{(x - \overline{x})^s}$$

EXAMPLE:

The numbers are 2, 3, 5, 9. The first moment for this set of numbers is

$$\bar{x} = \frac{2 + 3 + 5 + 9}{4} = 4.75$$

which is also the arithmetic mean.

The second moment is

$$\overline{x^2} = \frac{2^2 + 3^2 + 5^2 + 9^2}{4} = 29.75$$

The third moment is

$$\overline{x^3} = \frac{2^3 + 3^3 + 5^3 + 9^3}{4} = 222.25$$

The fourth moment is

$$\overline{x^4} = \frac{2^4 + 3^4 + 5^4 + 9^4}{4} = 1,820.75$$

EXAMPLE:

The measurements are 1, 3, 5, 15. The mean is $\bar{x} = 6$.

The first moment about the mean is

$$m_1 = \frac{(1 - 6) + (3 - 6) + (5 - 6) + (15 - 6)}{4} = 0$$

Note that the first moment about the mean is always equal to zero. Indeed

$$m_1 = \frac{\sum (x - \bar{x})}{n} = \frac{\sum x}{n} - \bar{x} = 0.$$

The second moment about the mean

$$m_2 = \overline{(x - \bar{x})^2}$$

$$= \frac{(1-6)^2 + (3-6)^2 + (5-6)^2 + (15-6)^2}{4} = 29$$

Observe that m_2 is the variance s^2.

The third moment about the mean

$$m_3 = \overline{(x - \bar{x})^3}$$

$$= \frac{(1-6)^3 + (3-6)^3 + (5-6)^3 + (15-6)^3}{4} = 220.5$$

We define the sth moment about any number A (called origin) as

$$m'_s = \frac{\sum\limits_{i=1}^{n} (x_i - A)^s}{n} = \overline{(x - A)^s}$$

We have

$$m'_s = \frac{\sum d_i^s}{n}$$

where $d_i = x_i - A$ is the deviation of x_i from A.

Similarly, the moments for grouped data can be computed. Let $x_1, x_2, \ldots, x_n$ be numbers which occur with frequencies $f_1, f_2, \ldots, f_n$ respectively.

The sth moment is defined by

61

$$\overline{x^S} = \frac{f_1 x_1{}^S + \ldots + f_n x_n{}^S}{f_1 + \ldots + f_n} = \frac{\sum f_i x_i{}^S}{\sum f_i}$$

The sth moment about the mean is defined by

$$m_S = \frac{\sum_{i=1}^{n} f_i (x_i - \overline{x})^S}{\sum_{i=1}^{n} f_i}$$

The sth moment about the origin A is defined as

$$m'_S = \frac{\sum_{i=1}^{n} f_i (x_i - A)^S}{\sum_{i=1}^{n} f_i}$$

There are some relations between moments. We shall list a few of them.

1.
$$m_2 = m'_2 - m'_1{}^2$$

$$m_2 = \overline{(x - \overline{x})^2} \quad \text{and} \quad d = x - A.$$

Then

$$x = d + A \quad \text{and} \quad \overline{x} = \overline{d} + A$$

$$x - \overline{x} = d - \overline{d}$$

We have

62

$$m_2 = \overline{(d - \overline{d})^2} = \overline{d^2 - 2\overline{d}d + \overline{d}^2}$$

$$= \overline{d^2} - 2\overline{d}\,\overline{d} + \overline{d}^2 = \overline{d^2} - \overline{d}^2$$

$$= m'_2 - m'^2_1$$

2.

$$m_3 = m'_3 - 3m'_1m'_2 + 2m'^3_1$$

$$m_3 = \overline{(x - \overline{x})^3} = \overline{(d - \overline{d})^3} = \overline{d^3 - 3d^2\overline{d} + 3d\overline{d}^2 - \overline{d}^3}$$

$$= \overline{d^3} - 3\overline{d^2}\,\overline{d} + 3\overline{d}^3 - \overline{d}^3 = \overline{d^3} - 3\overline{d}\,\overline{d^2} + 2\overline{d}^3$$

$$= m'_3 - 3m'_1m'_2 + 2m'^3_1$$

3. Similarly we prove

$$m_4 = m'_4 - 4m'_1m'_3 + 6m'^2_1 m'_2 - 3m'^4_1$$

and formulas for higher moments.

Any moment m_k can be expressed in terms of moments m_1', m_2', ..., m_k'.

Computation of Moments for Grouped Data

The coding method can be applied when the data x_i can be expressed in the form

$$x_i = A + cy_i$$

Equation

$$m'_s = \frac{\sum f_i(x_i - A)^s}{\sum f_i}$$

becomes

$$m'_s = c^s \frac{\sum f_i y_i^s}{\sum f_i} = c^s \overline{y^s}$$

That in turn enables us to compute moments m_s.

To verify the results we can use Charlier's check. The moments are calculated by the coding method and the identities are applied.

$$\sum f(y + 1) = \sum fy + \sum f$$

$$\sum f(y + 1)^2 = \sum fy^2 + 2\sum fy + \sum f$$

$$\sum f(y + 1)^3 = \sum fy^3 + 3\sum fy^2 + 3\sum fy + \sum f$$

$$\sum f(y + 1)^4 = \sum fy^4 + 4\sum fy^3 + 6\sum fy^2 + 4\sum fy + \sum f$$

Sometimes it is more convenient to use dimensionless units.

We define the dimensionless moments about the mean

$$P_r = \frac{m_r}{s_r} = \frac{m_r}{(\sqrt{m_2})^r} = \frac{m_r}{\sqrt{m_2^r}}$$

where $s = \sqrt{m_2}$ is the standard deviation.

5.2 COEFFICIENTS OF SKEWNESS – KURTOSIS

Distributions can be symmetric or asymmetric. For example, the normal distribution is symmetric. Among the asymmetric distributions, some can be "more" asymmetric than others. The degree of asymmetry is measured by skewness. Consider for example a distribution skewed to the right.

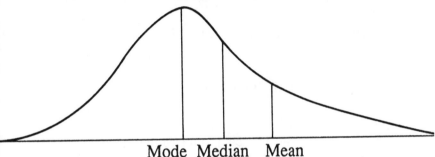

Mode   Median   Mean

The distribution skewed to the right has positive skewness. The asymmetry is measured by the difference.

Mean-Mode

To make it dimensionless we can divide Mean-Mode by a measure of dispersion, such as the standard deviation. Thus

$$\text{Skewness} = \frac{\text{Mean}-\text{Mode}}{\text{Standard Deviation}} = \frac{\bar{x} - \text{mode}}{s} \quad (*)$$

For a distribution skewed to the right, mean > mode and skewness is positive. The skewness of a distribution skewed to the left is negative. Using the empirical relation

65

$$\text{Mean - Mode} = 3 \text{ (Mean - Median)}$$

we can write

$$\boxed{\text{Skewness} = \frac{3(\text{Mean} - \text{Median})}{\text{Standard Deviation}} \quad (**)}$$

Equation (*) is called Pearson's first coefficient of skewness and equation (**) is called Pearson's second coefficient of skewness.

There are also other measures of skewness like

$$\text{Moment coefficient of skewness} = a_3 = \frac{m_3}{s^3} = \frac{m_3}{\sqrt{m_2^3}}$$

Also, $a_3{}^2 = \dfrac{m_3{}^2}{m_2{}^3}$ is used as a measure of skewness.

Skewness measures the degree of asymmetry. Kurtosis measures the shape of the peak of a distribution. Usually this is measured relative to a normal distribution. A distribution can have one of three kinds of peaks.

1. Leptokurtic, where a distribution has a relatively high peak.

2. Mesokurtic, where the peak is neither very high nor very low; the peak of the normal distribution, for example.

3. Platykurtic, where the distribution is flat and the peak is low and not sharply outlined.

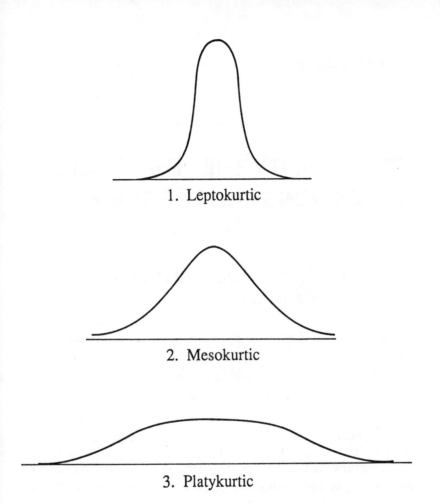

1. Leptokurtic

2. Mesokurtic

3. Platykurtic

Like skewness, kurtosis is described and measured in many ways. One measure of kurtosis is moment coefficient of kurtosis.

$$\text{Moment coefficient of kurtosis} = a_4 = \frac{m_4}{s^4} = \frac{m_4}{m_2^2}$$

For the normal distribution $a_4 = 3$.

CHAPTER 6

PROBABILITY THEORY – BASIC DEFINITIONS AND THEOREMS

6.1 CLASSICAL DEFINITION OF PROBABILITY

Let e denote an event which can happen in k ways out of a total of n ways. All n ways are equally likely. The probability of occurrence of the event e is defined as

$$p = pr\{e\} = \frac{k}{n}$$

The probability of occurrence of e is called its success. The probability of failure (non-occurrence) of the event is denoted by q.

$$q = pr\{\text{not } e\} = \frac{n-k}{n} = 1 - \frac{k}{n} = 1 - p$$

Hence, $p + q = 1$. The event "not e" is denoted by $\tilde{e}$ or $\sim e$.

EXAMPLE:

A toss of a coin will produce one of two possible outcomes {heads, tails}. Let e be the event that tails will turn up in a single toss of a coin. Then

$$p = \frac{1}{1+1} = \frac{1}{2}.$$

EXAMPLE:

We define the event e to be number 5 or 6 turning up in a single toss of a die. There are six equally likely outcomes of a single toss of a die.

$$\{1, 2, 3, 4, 5, 6\}$$

Thus, $n = 6$. The event e can occur in two ways:

$$p = pr\{e\} = \frac{2}{6} = \frac{1}{3}$$

Probability of failure of e is

$$q = pr\{\sim e\} = 1 - \frac{1}{3} = \frac{2}{3}.$$

For any event e

$$0 \leq pr\{e\} \leq 1$$

If the event cannot occur, its probability is 0. If the event must occur, its probability is 1.

Next we define the odds. Let p be the probability that an event will occur. The odds in favor of its occurrence are $p : q$ and the odds against it are $q : p$.

EXAMPLE:

We determine the probability that at least one tail appears in two tosses of a coin. Let h denote heads and t tails. The possible outcomes of two tosses are

$$(h, h), (h, t), (t, h), (t, t)$$

Three cases are favorable. Thus

$$p = \frac{3}{4}$$

EXAMPLE:

The event e is that the sum 8 appears in a single toss of a pair of dice. There are $6 \cdot 6 = 36$ outcomes:

$$(1, 1), (2, 1), (3, 1), \dots , (6, 6).$$

The sum 8 appears in five cases

$$(2, 6), (6, 2), (3, 5), (5, 3), (4, 4)$$

Then

$$p\{e\} = \frac{5}{36}$$

The concept of probability is based on the concept of random experiment. A random experiment is an experiment with more than one possible outcome, conducted such that it is not known in advance which outcome will occur. The set of possible outcomes is denoted by a capital letter, say, X. Usually, each outcome is either a number (a toss of a die) or something to which a number can be assigned (head = 1, tail = 0 for a toss of a coin).

For some experiments the number of possible outcomes is infinite.

6.2 RELATIVE FREQUENCY – LARGE NUMBERS

The classical definition of probability includes the assumption that all possible outcomes are equally likely. Often this is not the case. The statistical definition of probability is based on the notion of the relative frequency.

We define the statistical probability or empirical probability of an event as the relative frequency of occurrence of the event when the number of observations is very large. Then, the probability is the limit of the relative frequency as the number of observations increases indefinitely.

EXAMPLE:

Suppose a coin was tossed 1,000 times and the result was 587 tails. The relative frequency of tails is $\frac{587}{1000}$. Another 1,000 tosses lead to 511 tails. Then, the relative frequency of tails $= \frac{587 + 511}{1000 + 1000} = \frac{1098}{2000}$. Proceeding in this manner we obtain a sequence of numbers, which gets closer and closer to the number defined as the probability of a tail in a single toss.

The empirical probability is based on the principle called the Law of Large Numbers.

The Law of Large Numbers

The sample mean tends to approach the population mean.

Here, by an event, we understand a subset of possible outcomes. It may contain none, one, some, or all of the possible outcomes.

Now, we can define the probability as follows:

Definition of Probability

The probability of an event E is determined by associating 1 with the event occurring (success) and 0 with the event not occurring (failure). The experiment is performed a large number of times. The probability is defined as

$$\lim_{n \to \infty} \sum_{i=1}^{n} \frac{a_i}{n} = p = p(E)$$

where a_i is the outcome of the ith time the experiment is performed.

From the mathematical point of view this definition includes the concept of a limiting process, which may not exist. To avoid this trap, the axiomatic definition of probability was introduced.

6.3 INDEPENDENT AND DEPENDENT EVENTS: CONDITIONAL PROBABILITY

An event is a subset of all possible outcomes. Often, instead of saying event, we use the term set.

Union

The union of two sets, A and B, is the set of all elements which belong to A or to B. The union is denoted by $A \cup B$, read "A or B."

Intersection

The intersection of two sets, A and B, denoted by $A \cap B$, is the set containing all elements which belong to A and to B.

Difference

The difference of two sets, A and B, denoted by $A - B$, is the set of all elements of A which do not belong to B.

Complement

The complement of a set A, denoted by $\overline{A}$ or A^c or A', is the set of all elements (outcomes) in X which are not in A.

Subset

A is a subset of X, denoted $A \subset X$, if every element of A is an element of X.

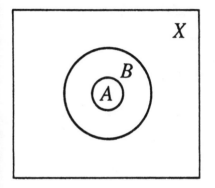

$A \subset B$

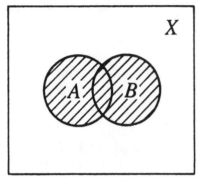

$A \cup B$ is shaded

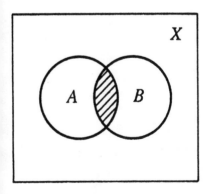

$A \cap B$ is shaded

$A - B$ is shaded

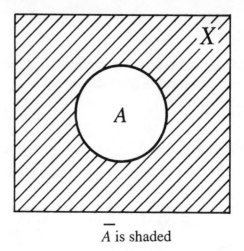

$\overline{A}$ is shaded

We will be using frequently the following identities

1. $A \cup B = B \cup A$, $A \cap B = B \cap A$

2. $A \cup (B \cup C) = (A \cup B) \cup C$
   $A \cap (B \cap C) = (A \cap B) \cap C$

3. $A \cup (B \cap C) = (A \cup B) \cap (A \cup C)$
   $A \cap (B \cup C) = (A \cap B) \cup (A \cap C)$

4. $\overline{\overline{A}} = A$

5. $A - B = A \cap \overline{B}$

DeMorgan's Laws

$$\overline{A \cup B} = \overline{A} \cap \overline{B}$$

$$\overline{A \cap B} = \overline{A} \cup \overline{B}$$

In most cases we assign an equal probability of $\frac{1}{n}$ to each of n possible outcomes. Sometimes it is difficult to determine the value of n.

Multiplication Principle

If one experiment has n possible outcomes and another has m possible outcomes, then the number of possible outcomes of performing first one experiment, then the other is

$$N = mn$$

Sampling with Replacement

Pick one of n balls from a bag and put it back. If the experiment is repeated m times, then the total number of possible outcomes is

$$N = n^m$$

Sampling without Replacement

Pick one of n balls from a bag and put it aside. Then pick another ball from the bag. Repeat this m times (where $m \leq n$). The number of possible outcomes is

$$N = n(n - 1)(n - 2) \ldots \cdot (n - m + 1) = \frac{n!}{(n - m)!}$$

Conditional Probability

The probability that E occurs given that F has occurred is denoted by

$$P(E \mid F)$$

or $P(E$ given $F)$ and is called the conditional probability of E given that F has occurred.

The event that "both E and F occur" is denoted by $E \cap F$ and is called a compound event.

We have

$$P(E \mid F) = \frac{P(E \cap F)}{P(F)}$$

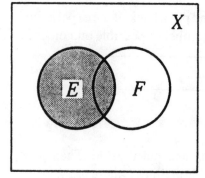

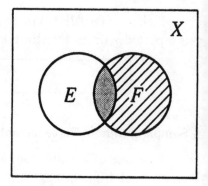

$P(E) =$	$P(E\mid F) =$
$\dfrac{\text{shaded}}{\text{shaded} + \text{unshaded}}$	$\dfrac{\text{shaded}}{\text{shaded} + \text{crosshatched}}$

Often we can find the probability of an intersection from

$$P(E \cap F) = P(E \mid F)\, P(F)$$

If

$$P(E \mid F) = P(E)$$

then we say that events E and F are independent, otherwise they are dependent. If E and F are independent then

$$P(E \cap F) = P(E) \cdot P(F)$$

For three events, E, F, G we have

$$P(E \cap F \cap G) = P(E) \, P(F \mid E) \, P(G \mid E \cap F)$$

If events E, F and G are independent then

$$P(E \cap F \cap G) = P(E) \cdot P(F) \cdot P(G)$$

One should not confuse independent events with mutually exclusive events. Two or more events are called mutually exclusive if the occurrence of any one of them excludes the occurrence of the others.

If E and F are independent, then

$$P(E \mid F) = P(E)$$

If E and F are mutually exclusive, then

$$P(E \mid F) = \frac{P(E \cap F)}{P(F)} = \frac{P(\emptyset)}{P(F)} = 0$$

Let $E_1, E_2, \ldots, E_n$ be a partition of the set Ω of all outcomes, i.e.,

$$E_i \cap E_j = \emptyset \ \text{ for } i \neq j$$

and

$$\bigcup_{i=1}^{n} E_i = \Omega$$

Then

$$P(E_1 \mid E) = \frac{P(E_1) \, P(E \mid E_1)}{\displaystyle\sum_{i=1}^{n} P(E_n) \, P(E \mid E_n)}$$

The last equation is called Bayes' Theorem.

6.4 THE CALCULUS OF PROBABILITY

We denote

$\emptyset$ = the set of no outcomes
Ω = the set of all possible outcomes (called certain event)

Here are the basic properties of probability:

$$P(\emptyset) = 0$$

$$P(\Omega) = 1$$

$$0 \leq P(E) \leq 1 \text{ for all events } E$$

$$P(E \cup F) = P(E) + P(F) - P(E \cap F)$$

If events E and F are mutually exclusive (i.e., $E \cap F = \emptyset$), then

$$P(E \cup F) = P(E) + P(F)$$

$$P(E \cup F \cup G) = P(E) + P(F) + P(G)$$

$$-P(E \cap F) - P(E \cap G) - P(F \cap G) + P(E \cap F \cap G)$$

In general $P(E_1 \cup E_2 \cup \ldots \cup E_n) = P(E_1) + \ldots + P(E_n)$
$- P$ (Intersection of twos) $+ P$ (Intersection of threes) $- \ldots \pm P(E_1$
$\cap \ldots \cap E_n)$.

$$P(\overline{E}) = 1 - P(E)$$

$$P(E \cap F) = P(E) + P(F) - P(E \cup F)$$

$$P(E \cap F) + P(E \cap \overline{F}) = P(E)$$

EXAMPLE:

The probability that A will be alive in ten years is 0.7 and the probability that B will be alive in ten years is 0.8. The probability that they both will be alive in ten years is

$$(0.7)\,(0.8) = 0.56$$

EXAMPLE:

A die is tossed twice.

E = event "1, 2, or 3" on the first toss.

F = event "1, 2, 3 or 4" on the second toss.

We compute the probability of getting a 1, 2, or 3 on the first toss and a 1, 2, 3 or 4 on the second toss.

$$P(E) = \frac{3}{6} = \frac{1}{2}$$

$$P(F) = \frac{4}{6} = \frac{2}{3}$$

Events E and F are independent. Hence

$$P(E \cap F) = \frac{1}{2} \cdot \frac{2}{3} = \frac{1}{3}$$

EXAMPLE:

The probability of a 5 turning up at least once in two tosses of a die.

E = event "5" on first toss.

F = event "5" on second toss.

$E \cup F$ = event "5" on first toss or "5" on second toss or both.

We shall compute $P(E \cup F)$. E and F are not mutually exclusive, hence

$$P(E \cup F) = P(E) + P(F) - P(E \cap F)$$

Since E and F are independent

$$P(E \cap F) = P(E) \cdot P(F)$$

we have

$$P(E \cup F) = P(E) + P(F) - P(E) \cdot P(F)$$

$$= \frac{1}{6} + \frac{1}{6} - \frac{1}{6} \cdot \frac{1}{6} = \frac{11}{36}.$$

EXAMPLE:

Two cards are drawn from a deck of 52 cards. Find the probability that they are both aces if the first card is

1. replaced

2. not replaced

E = event "ace" on first draw, and

F = event "ace" on second draw.

1. If the first card is replaced, E and F are independent, then

$$P(E \cap F) = P(E) P(F) = \frac{4}{52} \cdot \frac{4}{52} = \frac{1}{169}.$$

2. The first card is drawn and not replaced. Both cards can be drawn in $52 \cdot 51$ ways. E can occur in 4 ways and F in 3 ways. Then

$$P(E \cap F) = \frac{4 \cdot 3}{52 \cdot 51}$$

Another method:

$$P(E \cap F) = P(E)P(F \mid E) = \frac{4}{52} \cdot \frac{3}{51}$$

EXAMPLE:

Probability of drawing either an ace or a club or both from a deck of cards.

E = event "drawing an ace"

F = event "drawing a club"

Note that events E and F are not mutually exclusive. Then

$$P(E \cup F) = P(E) + P(F) - P(E \cap F)$$

$$= \frac{4}{52} + \frac{13}{52} - \frac{1}{52} = \frac{4}{13}$$

6.5 PROBABILITY DISTRIBUTIONS

Discrete Distributions

Variable X can assume a discrete set of values $x_1, x_2, \ldots, x_n$ with probabilities $p_1, p_2, \ldots, p_n$ respectively, where

$$p_1 + p_2 + \ldots + p_n = 1$$

This defines a discrete probability distribution for X.

Probability function or frequency function is defined by

$$p: x_i \to p_i \qquad i = 1, 2, \ldots, n$$

$$p(x_i) = p_i$$

Variable X, which assumes certain values with given probabilities, is called a discrete random variable.

EXAMPLE:

A pair of dice is tossed. X denotes the sum of the points obtained, $X = 2, 3, \ldots, 12$. The probability distribution is given by

x	2	3	4	5	6	7	8	9	10	11	12
$p(x)$	$\frac{1}{36}$	$\frac{2}{36}$	$\frac{3}{36}$	$\frac{4}{36}$	$\frac{5}{36}$	$\frac{6}{36}$	$\frac{5}{36}$	$\frac{4}{36}$	$\frac{3}{36}$	$\frac{2}{36}$	$\frac{1}{36}$

$$\sum p(x) = 1$$

Replacing probabilities with relative frequencies, we obtain from the probability distribution a relative frequency distribution. Probability distributions are for populations, while relative frequency distributions are for samples drawn from this population.

The probability distribution, like a relative frequency distribution, can be represented graphically.

Cumulative probability distributions are obtained by cumulating probabilities. The function describing this distribution is called a distribution function.

EXAMPLE:

Find the probability of boys and girls in families with four children. Probabilities for boys and girls are equal.

B = event "boy"

G = event "girl"

$$P(B) = P(G) = \frac{1}{2}$$

1. Four boys

$$P(B \cap B \cap B \cap B) = P(B) \cdot P(B) \cdot P(B) \cdot P(B) = \frac{1}{16}$$

2. Three boys and one girl

$$P(B \cap B \cap B \cap G \cup B \cap B \cap G \cap B$$

$$\cup B \cap G \cap B \cap B \cup G \cap B \cap B \cap B)$$

$$= P(B) \cdot P(B) \cdot P(B) \cdot P(G) \cdot 4$$

$$= \frac{1}{2} \cdot \frac{1}{2} \cdot \frac{1}{2} \cdot \frac{1}{2} \cdot 4 = \frac{1}{4}$$

3. Three girls and one boy same as above

$$p = \frac{1}{4}$$

4. Two boys and two girls

$$P(B \cap B \cap G \cap G \cup B \cap G \cap B \cap G \cup B \cap G \cap G \cap B$$

$$\cup G \cap G \cap B \cap B \cup G \cap B \cap G \cap B \cup G \cap B \cap B \cap G)$$

$$= P(B) \cdot P(B) \cdot P(G) \cdot P(G) \cdot 6 = \frac{6}{16} = \frac{3}{8}$$

5. Four girls

$$p = \frac{1}{16}$$

Number of Boys x	4	3	2	1	0
Probability $p(X)$	$\frac{1}{16}$	$\frac{4}{16}$	$\frac{6}{16}$	$\frac{4}{16}$	$\frac{1}{16}$

Here X is a random variable showing the number of boys in families with four children. The probability distribution is shown in the table.

This distribution can be represented graphically.

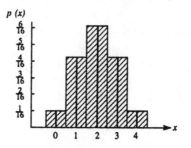

The sum of the areas of the rectangles is 1. Here the discrete variable X is treated as a continuous variable. The figure is called a probability histogram.

Continuous Distributions

Suppose variable X can assume a continuous set of values. In such a case, the relative frequency polygon of a sample becomes (or rather tends to) a continuous curve.

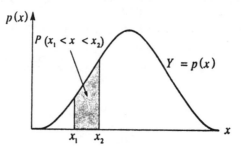

84

The total area under this curve is 1. The shaded area between the lines $x = x_1$ and $x = x_2$ is equal to the probability that x lies between x_1 and x_2.

Function $p(x)$ is a probability density function. Such a function defines a continuous probability distribution for X. The variable X is called a continuous random variable.

Mathematical Expectation

Let X be a discrete random variable which assumes the values $x_1, ..., x_n$ with respective probabilities $p_1, ..., p_n$ where $p_1 + p_2 + ... + p_n = 1$. The mathematical expectation of X denoted by $E(X)$ is defined as

$$E(X) = p_1 x_1 + ... + p_n x_n = \sum_{i=1}^{n} p_i x_i$$

6.6 COMBINATIONAL ANALYSIS

If an event can happen in any one of n ways and if, when this has occurred, another event can happen in any one of m ways, then the number of ways in which both events can happen in the specified order is

$$n \cdot m$$

EXAMPLE:

A bank offers three types of checking accounts and four types of savings accounts. Any customer who wants to have both accounts has $3 \cdot 4 = 12$ possibilities.

Factorial

Factorial n, denoted by $n!$, is defined as

$$n! = 1 \cdot 2 \cdot ... \cdot (n - 1) \cdot n \qquad 0! = 1$$

EXAMPLE:

$6! = 1 \cdot 2 \cdot 3 \cdot 4 \cdot 5 \cdot 6 = 720$

$1! = 1, \ 2! = 2$

EXAMPLE:

In how many ways can six children be arranged in a row?

Each of six children can fill the first position. Then each of five remaining children can fill the second place. Thus there are $6 \cdot 5$ ways of filling the first two places. There are four ways of filling the third place, three ways of filling the fourth place, two ways of filling the fifth place and one way of filling the sixth place. Hence

Number of arrangements $= 6 \cdot 5 \cdot 4 \cdot 3 \cdot 2 \cdot 1 = 6! = 720$

We have a general rule: number of arrangements of n different objects in a row is $n!$

Permutations

The number of ways k objects can be picked from among n objects with regard to order is denoted by $_nP_k$ or $P(n, k)$ or $P_{n,k}$ and is equal to

$$_nP_k = n(n - 1)\ (n - 2) \ldots (n - k + 1) = \frac{n!}{(n - k)!}$$

EXAMPLE:

Three letters are given x, y, z. There are $_3P_2$ ways of choosing two letters at a time:

$$_3P_2 = \frac{3!}{(3 - 2)!} = 2 \cdot 3 = 6$$

Indeed, there are six possibilities

$$xy, \ xz, \ yx, \ yz, \ zx, \ zy$$

EXAMPLE:

From the set of numbers $1, 2, 3, 4, 5, 6, 7, 8, 9$ one chooses three at a time. There are

$$_9P_3 = \frac{9!}{6!} = 7 \cdot 8 \cdot 9 = 504$$

three digit numbers.

Observe that the number of permutations of n objects, taken n at a time, is

$$_nP_n = \frac{n!}{(n-n)!} = n!$$

$_nP_k$ is called the number of permutations of n different objects, taken k at a time.

EXAMPLE:

In how many ways can ten people be arranged in a circle?

Person number one can be placed anywhere. The remaining nine people can be arranged in

$$9! = 362,880$$

ways.

The number of permutations of n objects consisting of groups in which n_1 are alike, n_2 are alike, ... is

$$\frac{n!}{n_1! \, n_2! \ldots}$$

where $n_1 + n_2 + \ldots = n$.

EXAMPLE:

The number of permutations of letters in the word criss-cross is

$$\frac{10!}{4! \, 2! \, 2! \, 1! \, 1!} = 37,800$$

Combinations

The number of ways one can choose k objects out of the n objects, disregarding the order, is denoted by

$$_nC_k, \; C(n,k) \text{ or } \binom{n}{k}$$

and is equal to

$$_nC_k = \frac{n(n+1)\ldots(n-k+1)}{k!} = \frac{n!}{k!\,(n-k)!} = \frac{_nP_k}{k!}$$

$_nC_k$ is the number of combinations of n objects taken k at a time.

EXAMPLE:

The number of combinations of the letters x, y, z taken two at a time is

$$_3C_2 = \frac{3!}{2!\,(3-2)!} = 3$$

The combinations are xy, xz, yz. Combinations xy and yx are the same, but permutations xy and yx are not the same.

EXAMPLE:

A committee consists of seven people. There are thirteen candidates. In how many ways can the committee be chosen?

$$_{13}C_7 = \frac{13!}{7! \, 6!} = \frac{8 \cdot 9 \cdot 10 \cdot 11 \cdot 12 \cdot 13}{2 \cdot 3 \cdot 4 \cdot 5 \cdot 6} = 1,716$$

EXAMPLE:

How many possible outcomes has a lotto game where a player chooses seven numbers out of forty-nine numbers?

$$_{49}C_7 = \frac{49!}{7! \, 42!} = \frac{43 \cdot 44 \cdot 45 \cdot 46 \cdot 47 \cdot 48 \cdot 49}{2 \cdot 3 \cdot 4 \cdot 5 \cdot 6 \cdot 7}$$

$$= 85,900,584$$

$$_nC_k = {}_nC_{n-k}$$

The number of combinations of n objects taken 1, 2, ..., n at a time is

$$_nC_1 + {}_nC_2 + ... + {}_nC_n = 2^n - 1$$

It is difficult to evaluate $n!$ for large numbers. In such cases an approximate formula (called Stirling's Formula) is used:

$$n! \approx \sqrt{2\pi n}\, n^n e^{-n}$$

where e is the natural base of logarithms, $e = 2.718281828...$

EXAMPLE:

Determine the probability of four 4's in six tosses of a die. The result of each toss is the event 4 or non 4 ($\overline{4}$). Thus

$$4, 4, 4, \overline{4}, 4, \overline{4}$$

or

$$\overline{4}, 4, 4, \overline{4}, 4, 4$$

are successes.

The probability of an event $4, 4, 4, \overline{4}, 4, \overline{4}$ is

$$P\left(4, 4, 4, \overline{4}, 4, \overline{4}\right) = \frac{1}{6} \cdot \frac{1}{6} \cdot \frac{1}{6} \cdot \frac{5}{6} \cdot \frac{1}{6} \cdot \frac{5}{6}$$

$$= \left(\frac{1}{6}\right)^4 \cdot \left(\frac{5}{6}\right)^2$$

All events in which four 4's and two non 4's occur have the same probability. The number of such events is

$$_6C_4 = \frac{6!}{4! \, 2!} = 15$$

and all these events are mutually exclusive. Hence

$$P(\text{four 4's in 6 tosses}) = 15 \cdot \left(\frac{1}{6}\right)^4 \cdot \left(\frac{5}{6}\right)^2 = 0.008$$

EXAMPLE:

30% of the cars produced by a factory have some defect. A sample of 100 cars is selected at random. What is the probability that

1. exactly 10 cars will be defective?

2. 95 or more will be defective?

In general, if $p = p(E)$ and $q = p(\overline{E})$ then the probability of getting exactly m E's in n trials is

$$_nC_m \, p^m q^{n-m}$$

1. $p(10 \text{ defective cars}) = {}_{100}C_{10} \left(\dfrac{3}{10}\right)^{10} \left(\dfrac{7}{10}\right)^{90}$

2. $p(95 \text{ or more defective}) = p(95 \text{ defective}) +$
   $p(96 \text{ defective}) + p(97 \text{ defective}) + p(98 \text{ defective}) +$
   $p(99 \text{ defective}) + p(100 \text{ defective}) =$

$$= {}_{100}C_{95} \left(\dfrac{3}{10}\right)^{95} \left(\dfrac{7}{10}\right)^{5} + {}_{100}C_{96} \left(\dfrac{3}{10}\right)^{96} \left(\dfrac{7}{10}\right)^{4}$$

$$+ \dots + {}_{100}C_{100} \left(\dfrac{3}{10}\right)^{100} \left(\dfrac{7}{10}\right)^{0}$$

6.7 SET THEORY AND PROBABILITY

First we define a sample space X, which consists of all possible outcomes of an experiment. With each point of X we can associate a non-negative number called a probability. If X contains only a finite number of points then the sum of all probabilities is equal to one.

An event is defined as a set of points in X.

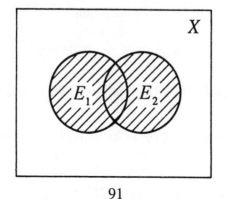

Next we define $E_1 \cup E_2$, $E_1 \cap E_2$, $E_1 - E_2$, and mutually exclusive events (i.e, events E_1 and E_2 such that $E_1 \cap E_2 = \emptyset$). The probability associated with the null set is zero:

$$p(\emptyset) = 0$$

EXAMPLE:

A die is tossed twice. Using a sample space we find the probability that the sum of two tosses is either three or seven. The sample space consists of all points $(1, 1)$, $(1, 2)$, …, $(6, 6)$.

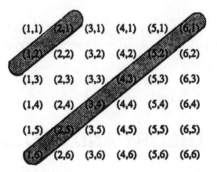

Sample space consists of thirty-six points. To each point we assign probability $\frac{1}{36}$:

$$36 \cdot \frac{1}{36} = 1$$

A = event "sum equals three"

B = event "sum equals seven"

$$p(A) = \frac{2}{36}, \quad p(B) = \frac{6}{36}$$

A and B have no points in common, that is, are mutually exclusive. Thus

$$p(A \cup B) = p(A) + p(B) = \frac{8}{36} = \frac{2}{9}.$$

CHAPTER 7

DISTRIBUTIONS

7.1 THE BINOMIAL DISTRIBUTION

Let E be an event and p be the probability that E will happen in any single trial. Number p is called the probability of a success. Then $q = 1 - p$ is the probability of a failure.

The probability that the event E will happen exactly x times in n trials is given by

$$p(x) = {}_nC_x p^x(1 - p)^{n-x} = \frac{n!}{x!(n-x)!}p^x(1-p)^{n-x} \quad (1)$$

$p(x)$ is the probability of x successes and $n - x$ failures.

The range of x is 0, 1, 2, ..., n.

Distribution (1) is called the Bernoulli distribution or the binomial distribution. For $x = 0, 1, ..., n$ we obtain in (1) ${}_nC_o, {}_nC_1, {}_nC_2, ..., {}_nC_n$ which are the binomial coefficients of the binomial expansion

$$(p + q)^n = {}_nC_o p^n + {}_nC_1 p^{n-1}q + {}_nC_2 p^{n-2}q^2 + ... + {}_nC_n q^n$$

EXAMPLE:

Toss a die five times. An event E is that a six appears. We find the binomial distribution.

$$p = \frac{1}{6}, \quad q = 1 - p = \frac{5}{6}, \quad n = 5$$

$$p(0) = {}_5C_0 \, p^0 q^5 = \left(\frac{5}{6}\right)^5 = 0.40187$$

$$p(1) = {}_5C_1 \left(\frac{1}{6}\right)\left(\frac{5}{6}\right)^4 = 0.40187$$

$$p(2) = {}_5C_2 \left(\frac{1}{6}\right)^2 \left(\frac{5}{6}\right)^3 = 0.16075$$

$$p(3) = {}_5C_3 \left(\frac{1}{6}\right)^3 \left(\frac{5}{6}\right)^2 = 0.03215$$

$$p(4) = {}_5C_4 \left(\frac{1}{6}\right)^4 \left(\frac{5}{6}\right)^1 = 0.00321$$

$$p(5) = {}_5C_5 \left(\frac{1}{6}\right)^5 \left(\frac{5}{6}\right)^0 = 0.00013$$

EXAMPLE:

Evaluate the expectation of x, i.e.,

$$\sum_{x=0}^{n} x p(x), \quad \text{for} \quad p(x) = {}_nC_x p^x q^{n-x}$$

$$\sum_{x=0}^{n} x p(x) = \sum_{x=1}^{n} x \, \frac{n!}{x!(n-1)!} \, p^x q^{n-x} \qquad (2)$$

$$= np \sum_{x=1}^{n} \frac{(n-1)!}{(x-1)!(n-x)!} p^{x-1} q^{n-x}$$

$$= np(p+q)^{n-1} = np$$

since $p + q = 1$.

EXAMPLE:

Evaluate the expectation of x^2, i.e., $\sum\limits_{x=0}^{n} x^2 p(x)$ where $p(x)$ is a binomial distribution.

$$\sum_{x=0}^{n} x^2 p(x) = \sum_{x=1}^{n} x^2 \frac{n!}{x!(n-x)!} p^x q^{n-x}$$

$$= \sum_{x=1}^{n} [x(x-1) + x] \frac{n!}{x!(n-x)!} p^x q^{n-x} \qquad (3)$$

$$= \sum_{x=2}^{n} x(x-1) \frac{n!}{x!(n-x)!} p^x q^{n-x}$$

$$+ \sum_{x=1}^{n} x \frac{n!}{x!(n-x)!} p^x q^{n-x}$$

$$= n(n-1) p^2 \sum_{x=2}^{n} \frac{(n-2)!}{(x-2)!(n-x)!} p^{x-2} q^{n-x} + np$$

$$\boxed{= n(n-1) p^2 (p+q)^{n-2} + np = n(n-1) p^2 + np}$$

Next we compute mean μ and variance σ^2 of a binomially distributed variable.

$$\mu = \sum_{x=0}^{n} x p(x) = np$$

$$\sigma^2 = \sum_{x=0}^{n} (x-\mu)^2 p(x) = \sum_{x=0}^{n} (x^2 - 2\mu x + \mu^2) p(x)$$

$$= \sum_{x=0}^{n} x^2 p(x) - 2\mu \sum_{x=0}^{n} x p(x) + \mu^2 \sum_{x=0}^{n} p(x)$$

$$= n(n-1) p^2 + np - 2\mu^2 + \mu^2$$

$$= np(1-p) = npq$$

95

EXAMPLE:

A coin is tossed 10,000 times. The mean number of heads is

$$\mu = np = 10,000 \; \frac{1}{2} = 5,000$$

In 10,000 tosses 5,000 heads are expected. The standard deviation is

$$\sigma = \sqrt{npq} = 100 \cdot \frac{1}{2} = 50$$

Properties of the Binomial Distribution

Mean	$\mu = np$
Standard deviation	$\sigma = \sqrt{npq}$
Variance	$\sigma^2 = npq$
Moment coefficient of skewness	$a_3 = \dfrac{q - p}{\sqrt{npq}}$
Moment coefficient of kurtosis	$a_4 = 3 + \dfrac{1 - 6pq}{npq}$

Observe that the binomial distribution is a sum of n distributions defined by

$$f(1) = p \qquad x \, \varepsilon \, \{0, 1\}$$
$$f(0) = q = 1 - p$$

Mean $\quad \mu = p$

Variance $\quad \sigma^2 = p(1 - p) = pq$

Some other distributions are also used, like: multinomial, hypergeometric, geometric and so on.

7.2 THE MULTINOMIAL DISTRIBUTION

The probability of events $E_1, E_2, ..., E_k$ are $p_1, p_2, ..., p_k$ respectively. The probability that $E_1, ..., E_k$ will occur $x_1, ..., x_k$ times respectively is

$$\frac{n!}{x_1!x_2!...x_k!} p_1^{x_1} p_2^{x_2} ... p_k^{x_k} \qquad (4)$$

where $x_1 + x_2 + ... + x_k = n$. Distribution (4) is called the multinomial distribution. The expected number of times each event $E_1, ..., E_k$ will occur in n trials are $np_1, ..., np_k$ respectively.

$$np_1 + ... + np_k = n(p_1 + ... + p_k) = n$$

Observe that (4) is the general term in the multinomial expansion $(p_1 + ... + p_k)^n$.

EXAMPLE:

A die is tossed ten times. The probability of getting 1, 2, 3, 4, and 5 exactly once and 6 five times is

$$\frac{10!}{5!}\left(\frac{1}{6}\right)\left(\frac{1}{6}\right)\left(\frac{1}{6}\right)\left(\frac{1}{6}\right)\left(\frac{1}{6}\right)\left(\frac{1}{6}\right)^5 = \frac{7 \cdot 8 \cdot 9 \cdot 10}{6^9}$$

$$= 0.0005$$

7.3 THE NORMAL DISTRIBUTION

The normal distribution is one of the most important examples of a continuous probability distribution. The equation

$$y = \frac{1}{\sigma\sqrt{2\pi}} e^{-\frac{(x-\mu)^2}{2\sigma^2}} \qquad (5)$$

is called a normal curve or Gaussian distribution. In (5) μ denotes mean and σ is the standard deviation. The total area bounded by (5) and the x–axis is one. Thus the area bounded by curve (5), the x-axis, and $x = a$ and $x = b$, where $a < b$, represents the probability that $x \, \varepsilon \, [a, b]$, denoted by

$$p(a < x < b)$$

A new variable z can be introduced by

$$z = \frac{x - \mu}{\sigma} \tag{6}$$

Then equation (5) becomes

$$y = \frac{1}{\sqrt{2\pi}} e^{-\frac{1}{2} z^2} \tag{7}$$

where $\sigma = 1$. Equation (7) is called the standard form of a normal distribution. Here, z is normally distributed with a mean of zero and a variance of one. The graph of the standardized normal curve is shown below.

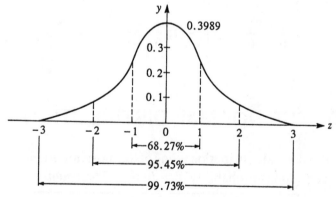

For example, the area between $z = -2$ and $z = 2$ under the curve is equal to 95.45% of the total area under the curve which is one.

We shall now list some properties of the normal distribution given by equation (5).

Mean	μ
Standard deviation	σ
Variance	σ^2
Moment coefficient of skewness (curve is symmetric)	$a_3 = 0$
Moment coefficient of kurtosis	$a_4 = 3$

EXAMPLE:

The weights of 300 men were measured. The mean weight was 160 lbs. and the standard deviation was 14 lbs. Assume that weights are normally distributed. We shall determine

1. how many men weigh more than 190 lbs.

2. how many men weigh between 145 and 165 lbs.

1. The weight of each man was rounded to the nearest pound. Men weighing more than 190 lbs. must weigh at least 190.5 lbs. In standard units

$$\frac{190.5 - 160}{14} = 2.18$$

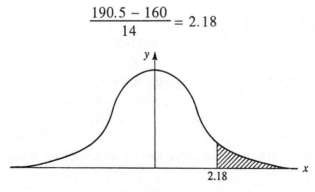

We have to find the shaded area.

Shaded Area = area to the right of 2.18
= (area to the right of $z = 0$) – (area between $z = 0$ and $z = 2.18$)
= 0.5 – 0.4854

$$= 0.0146$$

The number of men weighing more than 190 lbs. is

$$0.0146 \cdot 300 = 4.$$

2. Actually, we are interested in people weighing between 144.5 and 165.5 lbs. In standard units

$$\frac{144.5 - 160}{14} = -1.11 \qquad \frac{165.5 - 160}{14} = 0.39$$

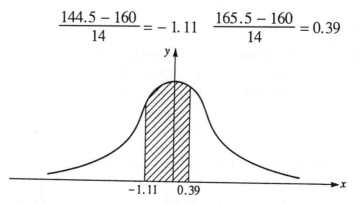

Area between $z = -1.11$ and $z = 0.39$
= (area between z –1.11 and $z = 0$) + (area between $z = 0$ and $z = 0.39$)
= 0.3665 + 0.1517

$$=0.5182$$

The number of men weighing between 145 and 165 lbs. is

$$0.5182 \cdot 300 = 155.$$

Note that the areas of interest were obtained from the tables for the standard normal curve. In terms of probability we can write the results

$$p\,(w \geq 189.5) = 0.0146$$

$$p\,(\,144.5 \leq w \leq 165.5) = 0.5182$$

For number n large, and p and q which are not too close to zero, the binomial distribution can be fairly well approximated by a normal distribution with the standardized variable

$$z = \frac{x - np}{\sqrt{npq}}$$

With increasing n the approximation becomes better, and for $n \to \infty$ it becomes a normal distribution.

7.4 THE POISSON DISTRIBUTION

The Poisson distribution is the discrete probability distribution defined by

$$p(x) = \frac{\lambda^x e^{-\lambda}}{x!} \qquad (8)$$

where $x = 0, 1, 2, \ldots$
e and λ are constants.

Here are some properties of the Poisson distribution.

Mean	$\mu = \lambda$
Standard deviation	$\sigma = \sqrt{\lambda}$
Variance	$\sigma = \lambda$

Moment coefficient of skewness	$a_3 = \dfrac{1}{\sqrt{\lambda}}$
Moment coefficient of kurtosis	$a_4 = 3 + \dfrac{1}{\lambda}$

The Poisson distribution approaches a normal distribution with the standardized variable $\dfrac{x - \lambda}{\sqrt{\lambda}}$ as λ increases to infinity.

Under certain conditions the binomial distribution is very closely approximated by the Poisson distribution. These conditions are: n is very large, the probability p of occurrence of an event is close to zero (such an event is called a rare event), and $q = 1 - p$ is close to one.

In applications, an event is considered rare if n is at least $50(n \geq 50)$ while $np < 5$. Then the binomial distribution is closely approximated by the Poisson distribution with $\lambda = np$.

EXAMPLE:

Five percent of people have high blood pressure. Find the probability that in a sample of ten people chosen at random, exactly three will have high blood pressure.

First we use the binomial distribution:

$$p(\text{high blood pressure}) = \frac{1}{20}$$

$$p(3 \text{ in } 10) = {}_{10}C_3 \left(\frac{1}{20}\right)^3 \left(1 - \frac{1}{20}\right)^7$$

$$= \frac{10!}{7!3!}\left(\frac{1}{20}\right)^3 \left(\frac{19}{20}\right)^7 = 0.01$$

We solve this problem using the Poisson approximation to the binomial distribution

$$p(3 \text{ in } 10) = \frac{\lambda^x e^{-\lambda}}{x!}$$

We set $\lambda = np = 10 \cdot \dfrac{1}{20} = 0.5$

$$p = \frac{(0.5)^3 e^{-0.5}}{3!} = 0.01$$

In this case $p = \dfrac{1}{20}$ (close to zero) and $np < 5$. approximation is good.

"The ESSENTIALS" of Math & Science

Each book in the ESSENTIALS series offers all essential information of the field it covers. It summarizes what every textbook in the particular field must include, and is designed to help students in preparing for exams and doing homework. The ESSENTIALS are excellent supplements to any class text.

The ESSENTIALS are complete and concise with quick access to needed information. They serve as a handy reference source at all times. The ESSENTIALS are prepared with REA's customary concern for high professional quality and student needs.

Available in the following titles:

Advanced Calculus I & II
Algebra & Trigonometry I & II
Anatomy & Physiology
Anthropology
Astronomy
Automatic Control Systems /
   Robotics I & II
Biology I & II
Boolean Algebra
Calculus I, II, & III
Chemistry
Complex Variables I & II
Computer Science I & II
Data Structures I & II
Differential Equations I & II
Electric Circuits I & II
Electromagnetics I & II

Electronics I & II
Electronic Communications I & II
Fluid Mechanics /
   Dynamics I & II
Fourier Analysis
Geometry I & II
Group Theory I & II
Heat Transfer I & II
LaPlace Transforms
Linear Algebra
Math for Computer Applications
Math for Engineers I & II
Math Made Nice-n-Easy I to XII
Mechanics I, II, & III
Microbiology
Modern Algebra
Molecular Structures of Life

Numerical Analysis I & II
Organic Chemistry I & II
Physical Chemistry I & II
Physics I & II
Pre-Calculus
Probability
Psychology I & II
Real Variables
Set Theory
Sociology
Statistics I & II
Strength of Materials &
   Mechanics of Solids I &
Thermodynamics I & II
Topology
Transport Phenomena I &
Vector Analysis

If you would like more information about any of these books,
complete the coupon below and return it to us or visit your local bookstore.

RESEARCH & EDUCATION ASSOCIATION
61 Ethel Road W. • Piscataway, New Jersey 08854
Phone: (732) 819-8880   **website: www.rea.com**

Please send me more information about your Math & Science Essentials books

Name _____

Address _____

City _____ State _____ Zip _____

REA's **Problem Solvers**

The "PROBLEM SOLVERS" are comprehensive supplemental textbooks designed to save time in finding solutions to problems. Each "PROBLEM SOLVER" is the first of its kind ever produced in its field. It is the product of a massive effort to illustrate almost any imaginable problem in exceptional depth, detail, and clarity. Each problem is worked out in detail with a step-by-step solution, and the problems are arranged in order of complexity from elementary to advanced. Each book is fully indexed for locating problems rapidly.

ACCOUNTING	LINEAR ALGEBRA
ADVANCED CALCULUS	MACHINE DESIGN
ALGEBRA & TRIGONOMETRY	MATHEMATICS for ENGINEERS
AUTOMATIC CONTROL	MECHANICS
SYSTEMS/ROBOTICS	NUMERICAL ANALYSIS
BIOLOGY	OPERATIONS RESEARCH
BUSINESS, ACCOUNTING, & FINANCE	OPTICS
CALCULUS	ORGANIC CHEMISTRY
CHEMISTRY	PHYSICAL CHEMISTRY
COMPLEX VARIABLES	PHYSICS
DIFFERENTIAL EQUATIONS	PRE-CALCULUS
ECONOMICS	PROBABILITY
ELECTRICAL MACHINES	PSYCHOLOGY
ELECTRIC CIRCUITS	STATISTICS
ELECTROMAGNETICS	STRENGTH OF MATERIALS &
ELECTRONIC COMMUNICATIONS	MECHANICS OF SOLIDS
ELECTRONICS	TECHNICAL DESIGN GRAPHICS
FINITE & DISCRETE MATH	THERMODYNAMICS
FLUID MECHANICS/DYNAMICS	TOPOLOGY
GENETICS	TRANSPORT PHENOMENA
GEOMETRY	VECTOR ANALYSIS
HEAT TRANSFER	

*If you would like more information about any of these books,
complete the coupon below and return it to us or visit your local bookstore.*

RESEARCH & EDUCATION ASSOCIATION
61 Ethel Road W. • Piscataway, New Jersey 08854
Phone: (732) 819-8880   **website: www.rea.com**

Please send me more information about your Problem Solver books

Name _____

Address ~~DISCARDED~~ _____

City _____ State _____ Zip _____